SpringerBriefs in Computer Science

More information about this series at http://www.springer.com/series/10028

Hamidreza Alvari • Elham Shaabani
Paulo Shakarian

Identification of Pathogenic Social Media Accounts

From Data to Intelligence to Prediction

 Springer

Hamidreza Alvari
Fulton Schools of Engineering, CIDSE
Arizona State University
Tempe, AZ, USA

Elham Shaabani
Fulton Schools of Engineering, CIDSE
Arizona State University
Tempe, AZ, USA

Paulo Shakarian
Fulton Schools of Engineering, CIDSE
Arizona State University
Tempe, AZ, USA

ISSN 2191-5768   ISSN 2191-5776   (electronic)
SpringerBriefs in Computer Science
ISBN 978-3-030-61430-0   ISBN 978-3-030-61431-7   (eBook)
https://doi.org/10.1007/978-3-030-61431-7

This Springer imprint is published by the registered company Springer Nature Switzerland AG
The registered company address is: Gewerbestrasse 11, 6330 Cham, Switzerland

Acknowledgments

The authors would like to acknowledge the support from the Department of Defense (Minerva Program), US State Department Global Engagement Center, National Science Foundation, Army Research Office (including grants W911NF1910066 and W911NF1510282), and Office of Naval Research. We would also like to thank the following collaborators for their contributions in some of the research discussed in the book: Ghazaleh Beigi, Ashkan Sadeghi-Mobarakeh, Soumajyoti Sarkar, Ruocheng Guo, Scott W. Ruston, Steven R. Corman, and Hasan Davulcu.

Contents

Chapter 1
Introduction

Recent years have witnessed an exponential growth of online platforms such as online social networks (OSNs) and microblogging websites. These platforms play a major role in online communication and information sharing as they have become large-scale and real-time communication tools. This leads to massive user-generated data produced on a daily basis and via different forms that are rich sources of information and can be used in different tasks from marketing to research. On the negative side, online platforms have become widespread tools exploited by various malicious actors who orchestrate societal-significant threats leading to numerous security and privacy issues [4, 6, 10–15, 18, 19, 23, 44].

To better understand the behavior and impact of malicious actors and counter their activity, social media and other online platforms' authorities need to deploy certain capabilities to reduce their threats. Due to the large volume of information published online and because of the limited manpower, the burden usually falls to algorithms that are designed to automatically identify these bad actors. This is a subtle task facing online platforms due to several challenges: (1) malicious users have strong incentives to disguise themselves as normal users (e.g., intentional misspellings, camouflaging, etc.), (2) malicious users are highly likely *key* users in making harmful messages go viral and thus need to be detected at their early life span to stop their threats from reaching a vast audience, and (3) available data for training automatic approaches for detecting malicious users are usually either highly imbalanced (i.e., higher number of normal users than malicious users) or comprise insufficient labeled data.

In this book, we present the results of a research program that focuses on investigating malicious information cascades formed by resharing mechanisms in social media attributed to "Pathogenic Social Media" (PSM). In particular, we set out to understand PSM accounts who are believed to be key users in making malicious campaigns. Various machine learning-based algorithms are then presented to detect PSM accounts ranging from supervised to semi-supervised learning algorithms. Below, we first briefly explain the problem of identification of PSM accounts and then present research challenges. We then summarize the contributions of this book and finally conclude this chapter by reviewing the literature related to the results presented in this book.

Resharing mechanisms on social media such as retweeting in Twitter allow massive spread of harmful disinformation to viral proportions. Manipulating public opinion and political events on the Web can be attributed to accounts dedicated to spreading malicious information, referred to as PSM accounts (e.g., terrorist supporters, or fake news writers) [5]. PSMs are users who seek to promote or degrade certain ideas by utilizing large online communities of supporters to reach their goals. Identifying PSMs has immediate applications including countering terrorism [34, 35], fake news detection [30, 31], and water armies detection [20].

Detecting PSMs in social media is crucial as they are likely key users in formation of malicious campaigns [49]. This is a challenging task for several reasons. First, these platforms are primarily based on reports they receive from their own users[1] to manually shut down PSMs. This straightforward solution is not necessarily a timely approach since despite efforts to suspend these accounts, many of them simply return to social media with different accounts which makes their manual suspension a non-trivial task. Second, available data for training automatic PSM detection approaches is often imbalanced and social network structure, which is at the core of many techniques [2, 3, 9, 16, 33, 50, 54], is not readily available. Third, PSMs often seek to utilize and cultivate large number of online communities of passive supporters to spread as much harmful information as they can while disguising themselves as normal users. To address the aforementioned challenges, we review several methods and algorithms to detect PSM accounts in their early life span.

This book addresses the following challenges facing online platforms in identifying PSM accounts:

- PSM users are highly likely key users in making harmful messages go *"viral"*— where "viral" is defined as an order-of-magnitude increase. Mechanisms are thus required to stop their threats from reaching vast audience early enough to stop formation of malicious campaigns [5]. Throughout this book, probabilistic causal inference [37] is tailored to identify PSMs since they are key users in making a harmful message viral. Chapter 2 seeks to distinguish PSM from non-PSM users by utilizing causal inference and Hawkes processes [1]. In Chaps. 3 and 4, standard and time-decay causal metrics are utilized to distinguish PSMs from normal users within a *short* time around their activity [5, 42].
- PSM users often seek to utilize and cultivate large number of online communities and campaigns of passive supporters to spread as much harmful information as they can. Consequently, in Chap. 4, we investigate whether or not causality scores of PSM users within same communities are higher than those across different communities. Accordingly, a causal community detection-based classification

---

[1] https://bit.ly/2Dq5i4M.

method is proposed that takes causality-based attribute vectors of users and the community structure of their action log to detect them.

- Available data for training automatic approaches for detecting PSM users are usually either highly imbalanced (i.e., higher number of normal users than malicious users) or comprise insufficient labeled data [5]. To overcome the issue of lack of enough annotated data, Chaps. 5 and 6 propose semi-supervised approaches for detecting PSMs [7, 43]. More specifically, Chap. 5 leverages causal inference and Laplacian semi-supervised SVM to detect PSM users, and Chap. 6 proposes a combined approach of graph-based and causal metrics via supervised and semi-supervised settings.
- PSM users have strong incentives to disguise themselves as normal users (e.g., camouflaging [32]). This makes the task of malicious users identification a daunting task. Later in Chap. 7, we observe that PSMs would deploy techniques to generate diverse information to make their posts look more natural. We utilize several metrics to distinguish content shared by PSM accounts and normal users.
- Despite PSM users' effort to act normally, they still behave significantly differently than normal users on many levels [51]. In Chap. 7, we take a closer look at the differences between malicious and normal behavior in terms of their posted URLs.

The research covered in this book is related to a number of other research directions. Below, we will summarize some of the state-of-the-art methods in each category while highlighting their differences with our work.

Social Spam/Bot Detection The closest research direction to our research is social spam/bot detection. Previous work on spam/bot detection either assumed network information is available [28] or did not differentiate between types of bots [25]. However, we did not assume any underlying network structure or cascade path information, and our approach is specific to PSM accounts. For example, recently, DARPA organized a Twitter bot challenge to detect "influence bots" [46]. Among the participants, the work of [19] used similarity to cluster accounts and uncover groups of malicious users. The work of [48] presented a supervised framework for bot detection which uses more than thousands features. In a different attempt, the work of [29] studied the problem of spam detection in Wikipedia using different spammers behavioral features. There also exist some studies in the literature that have addressed (1) differences between humans and bots [22], (2) different natures of bots [48], or (3) differences between bots and human trolls [18]. For example, the work of [22] conducted a series of measurements in order to distinguish humans from bots and cyborgs, in term of tweeting behavior, content, and account properties. To do so, they used more than 40 million tweets posted by over 500 K users. Then, they performed analysis and find groups of features that are useful for classifying users into human, bots, and cyborgs. They concluded that entropy and certain account properties can be very helpful in differentiating between those accounts. In a different attempt, some other studies have tried to differentiate between several natures of bots. For instance, in the work of [48], the authors performed clustering analysis and revealed specific behavioral groups of accounts. Specifically, they identified different types of bots such as *spammers*, *self-promoters*, and *accounts that post content from connected applications*, using manual investigation of samples extracted from clusters. Their cluster analysis emphasized that Twitter hosts a variety of users with diverse behaviors; that is, in some cases the boundary between human and bot users is not sharp, i.e., some account exhibit characteristics of both.

Also, the work of [18] uses Twitter data to quantify the impact of Russian trolls and bots on amplifying polarizing and anti-vaccine tweets. They first used the Botometer API to assign bot probabilities to the users in the dataset and divided the whole dataset into 3 categories: those with scores less than 20% (very likely to be human), between 20% and 80% (e.g., cyborgs with uncertain provenance) and above 80% (high likely to be bots). Then, they posed two research questions: (1) Are bots and trolls more likely to tweet about vaccines? and (2) Are bots and trolls more likely to tweet polarizing and anti-vaccine content? Their analysis demonstrated that Twitter bots and trolls significantly impact on online discussion about vaccination and this differs by account type. For example, Russian trolls and bots post content about vaccination at higher rates compared to an average user. Also, according to this study, troll accounts and content polluters (e.g., dissemination of malware, unsolicited commercial content, etc.) post anti-vaccine tweets 75% more than average users. In contrast, spambots, which can be easily distinguished from humans, are less likely to promote anti-vaccine messages. Their closing remarks suggest strongly that distinguishing between malicious actors (bots, trolls, cyborgs, and human users) is difficult and thus anti-vaccine messages may be disseminated at higher rates by a combination of these malicious actors.

Fake News Identification A growing body of research has addressed the impact of bots in manipulating political discussion, including the 2016 U.S. presidential election [44] and the 2017 French election [26]. For example, [44] analyzed tweets following recent U.S. presidential election and found evidences that bots played key roles in spreading fake news.

Identifying Instigators Given a snapshot of the diffusion process at a given time, these works aim to detect the source of the diffusion. For instance, [56] designed an approach for information source detection and in particular initiator of a cascade. In contrast, we are focused on a set of users who *might* or *might not* be initiators. Other similar works on finding most influential spreaders of information such as [27, 39] and outbreak prediction such as [23] also exist in the literature. For example, the work of [38] performed classification to detect users who adopt popular items. In [56], the authors designed an approach for information source detection and in particular initiator of a cascade.

Extremism and Water Armies Detection Several studies have focused on understanding extremism in social networks [6, 17, 35, 40, 41]. The work of [35] used Twitter and proposed an approach to predict new extremists, determine if the newly created account belongs to a suspended extremist, and predict the ego-network of the suspended extremist upon creating her new account. The authors in [17] performed iterative vertex clustering and classification to identify Islamic Jihadists on Twitter. The term "Internet water armies" refers to a special group of online users who get paid for posting comments for some hidden purposes such as influencing other users towards social events or business markets. Therefore, they are also called "hidden paid posters." The works of [20, 21] used user behavioral and domain-specific attributes and designed approaches to detect Internet water armies. The work of [20] also used user behavioral and domain-specific attributes to detect water armies.

Causal Reasoning As opposed to [36, 37, 45] which dealt with preconditions as single atomic propositions, in this book, we use rules with preconditions of more than one atomic propositions.

Point Processes When dealing with timestamped events in continuous time such as the activity of users on social media, point processes could be leveraged for modeling such events. Point processes have been extensively used to model activities in networks [52]. Hawkes processes are a special form of point processes which models complicated event sequences with historical events influencing future ones. Hawkes processes have been applied to a variety of problems including financial analysis [8], seismic analysis [24] and social network modeling [55], community detection [47], and causal inference [53].

References

1. H. Alvari, P. Shakarian, Hawkes process for understanding the influence of pathogenic social media accounts, in *2019 2nd International Conference on Data Intelligence and Security (ICDIS)*, pp. 36–42, June 2019
2. H. Alvari, S. Hashemi, A. Hamzeh, Detecting overlapping communities in social networks by game theory and structural equivalence concept, in *International Conference on Artificial Intelligence and Computational Intelligence* (Springer, 2011), pp. 620–630
3. H. Alvari, A. Hajibagheri, G. Sukthankar, K. Lakkaraju, Identifying community structures in dynamic networks. Soc. Netw. Anal. Min. **6**(1), 77 (2016)
4. H. Alvari, P. Shakarian, J. Snyder, A non-parametric learning approach to identify online human trafficking, in *2016 IEEE Conference on Intelligence and Security Informatics (ISI)*, pp. 133–138 (2016)
5. H. Alvari, E. Shaabani, P. Shakarian, Early identification of pathogenic social media accounts. *IEEE Intelligent and Security Informatics* (2018). arXiv:1809.09331
6. H. Alvari, S. Sarkar, P. Shakarian, Detection of violent extremists in social media, in *IEEE Conference on Data Intelligence and Security* (2019)
7. H. Alvari, E. Shaabani, S. Sarkar, G. Beigi, P. Shakarian, Less is more: Semi-supervised causal inference for detecting pathogenic users in social media, in *Companion Proceedings of The 2019 World Wide Web Conference* (ACM, 2019), pp. 154–161
8. E. Bacry, T. Jaisson, J.-F. Muzy, Estimation of slowly decreasing Hawkes kernels: application to high-frequency order book dynamics. Quantitative Finance **16**(8), 1179–1201 (2016)
9. G. Beigi, H. Liu, Similar but different: Exploiting users' congruity for recommendation systems, in *International Conference on Social Computing, Behavioral-Cultural Modeling and Prediction and Behavior Representation in Modeling and Simulation* (Springer, 2018), pp. 129–140

10. G. Beigi, H. Liu, A survey on privacy in social media: Identification, mitigation, and applications. ACM Trans. Data Sci. **1**(1), 1–38 (2020)
11. G. Beigi, M. Jalili, H. Alvari, G. Sukthankar, Leveraging community detection for accurate trust prediction, in *In ASE International Conference on Social Computing, Palo Alto, CA* (May 2014)
12. G. Beigi, K. Shu, Y. Zhang, H. Liu, Securing social media user data-an adversarial approach, in *Proceedings of the 29th on Hypertext and Social Media*, pp. 156–173 (2018)
13. G. Beigi, R. Guo, A. Nou, Y. Zhang, H. Liu, Protecting user privacy: An approach for untraceable web browsing history and unambiguous user profiles, in *Proceedings of the Twelfth ACM International Conference on Web Search and Data Mining* (ACM, 2019), pp. 213–221
14. G. Beigi, K. Shu, R. Guo, S. Wang, H. Liu, Privacy preserving text representation learning, in *Proceedings of the 30th ACM Conference on Hypertext and Social Media*, pp. 275–276 (2019)
15. G. Beigi, A. Mosallanezhad, R. Guo, H. Alvari, A. Nou, H. Liu, Privacy-aware recommendation with private-attribute protection using adversarial learning, in *Proceedings of the Thirteenth ACM International Conference on Web Search and Data Mining* (ACM, 2020)
16. G. Beigi, J. Tang, H. Liu, Social science–guided feature engineering: A novel approach to signed link analysis. ACM Trans. Intell. Syst. Technol. **11**(1), 1–27 (Jan. 2020)
17. M.C. Benigni, K. Joseph, K.M. Carley, Online extremism and the communities that sustain it: Detecting the isis supporting community on twitter. PloS one (2017). https://doi.org/10.1371/journal.pone.0181405
18. D.A. Broniatowski, A.M. Jamison, S. Qi, L. AlKulaib, T. Chen, A. Benton, S.C. Quinn, M. Dredze, Weaponized health communication: Twitter bots and Russian trolls amplify the vaccine debate. Am. J. Public Health **108**(10), 1378–1384 (2018)
19. Q. Cao, X. Yang, J. Yu, C. Palow, Uncovering large groups of active malicious accounts in online social networks, in *CCS* (2014)
20. C. Chen, K. Wu, S. Venkatesh, X. Zhang, Battling the internet water army: Detection of hidden paid posters. CoRR, abs/1111.4297 (2011)
21. C. Chen, K. Wu, S. Venkatesh, R.K. Bharadwaj, The best answers? think twice: online detection of commercial campaigns in the CQA forums, in *ASONAM* (2013)
22. Z. Chu, S. Gianvecchio, H. Wang, S. Jajodia, Detecting automation of twitter accounts: Are you a human, bot, or cyborg? IEEE Trans. Dependable Secure Comput. **9**(6), 811–824 (2012)
23. P. Cui, S. Jin, L. Yu, F. Wang, W. Zhu, S. Yang, Cascading outbreak prediction in networks: A data-driven approach, in *KDD* (2013)
24. D.J. Daley, D. Vere-Jones, *An Introduction to the Theory of Point Processes: Volume II: General Theory and Structure* (Springer Science & Business Media, 2007)
25. J.P. Dickerson, V. Kagan, V.S. Subrahmanian, Using sentiment to detect bots on twitter: Are humans more opinionated than bots? in *ASONAM* (2014)
26. E. Ferrara, Disinformation and social bot operations in the run up to the 2017 French presidential election (2017)
27. H.C.-Y. Fu, Yu-Hsiang, C.-T. Sun, Identifying Super-Spreader Nodes in Complex Networks, Math. Probl. Eng. vol. 2015, Article ID 675713, Page 8, (2015). https://doi.org/10.1155/2015/675713
28. A. Goyal, F. Bonchi, L.V. Lakshmanan, Learning influence probabilities in social networks, in *WSDM* (2010)
29. T. Green, F. Spezzano, Spam users identification in Wikipedia via editing behavior, in *ICWSM* (2017)
30. A. Gupta, H. Lamba, P. Kumaraguru, $1.00 per rt #bostonmarathon #prayforboston: Analyzing fake content on twitter, in *2013 APWG eCrime Researchers Summit* (2013)
31. A. Gupta, P. Kumaraguru, C. Castillo, P. Meier, *TweetCred: Real-Time Credibility Assessment of Content on Twitter* (Springer International Publishing, 2014)
32. B. Hooi, H.A. Song, A. Beutel, N. Shah, K. Shin, C. Faloutsos, Fraudar: Bounding graph fraud in the face of camouflage, in *Proceedings of the 22nd ACM SIGKDD International Conference on Knowledge Discovery and Data Mining* (ACM, 2016), pp. 895–904

33. D. Kempe, J. Kleinberg, E. Tardos, Maximizing the spread of influence through a social network, in *KDD* (2003)
34. M. Khader, *Combating Violent Extremism and Radicalization in the Digital Era*. Advances in Religious and Cultural Studies (IGI Global, 2016)
35. J. Klausen, C. Marks, T. Zaman, Finding online extremists in social networks. CoRR, abs/1610.06242 (2016)
36. S. Kleinberg, A logic for causal inference in time series with discrete and continuous variables, in *IJCAI* (2011)
37. S. Kleinberg, B. Mishra, The temporal logic of causal structures. CoRR, abs/1205.2634 (2012)
38. T. Konishi, T. Iwata, K. Hayashi, K.-I. Kawarabayashi, Identifying key observers to find popular information in advance, in *IJCAI* (2016)
39. S. Pei, L. Muchnik, J.S.A. Jr., Z. Zheng, H.A. Makse, Searching for superspreaders of information in real-world social media. CoRR (2014)
40. J.R. Scanlon, M.S. Gerber, Automatic detection of cyber-recruitment by violent extremists. Security Informatics **3**(1), 5 (2014)
41. J.R. Scanlon, M.S. Gerber, Forecasting violent extremist cyber recruitment. IEEE Trans. Inf. Forensics Sec. **10**(11), 2461–2470 (2015)
42. E. Shaabani, R. Guo, P. Shakarian, Detecting pathogenic social media accounts without content or network structure, in *2018 1st International Conference on Data Intelligence and Security (ICDIS)* (IEEE, 2018), pp. 57–64
43. E. Shaabani, A. Sadeghi-Mobarakeh, H. Alvari, P. Shakarian, An end-to-end framework to identify pathogenic social media accounts on twitter, in *IEEE Conference on Data Intelligence and Security* (2019)
44. C. Shao, G.L. Ciampaglia, O. Varol, A. Flammini, F. Menczer, The spread of fake news by social bots. Preprint (2017). arXiv:1707.07592
45. A. Stanton, A. Thart, A. Jain, P. Vyas, A. Chatterjee, P. Shakarian, Mining for causal relationships: A data-driven study of the Islamic state. CoRR (2015)
46. V.S. Subrahmanian, A. Azaria, S. Durst, V. Kagan, A. Galstyan, K. Lerman, L. Zhu, E. Ferrara, A. Flammini, F. Menczer, The DARPA twitter bot challenge (2016)
47. L. Tran, M. Farajtabar, L. Song, H. Zha, NetCodec: Community detection from individual activities, in *Proceedings of the 2015 SIAM International Conference on Data Mining* (SIAM, 2015), pp. 91–99
48. O. Varol, E. Ferrara, C.A. Davis, F. Menczer, A. Flammini, Online human-bot interactions: Detection, estimation, and characterization, in *ICWSM* (2017)
49. O. Varol, E. Ferrara, F. Menczer, A. Flammini, Early detection of promoted campaigns on social media. EPJ Data Sci. 6, 13 (2017)
50. L. Weng, F. Menczer, Y.-Y. Ahn, Predicting successful memes using network and community structure, in *ICWSM* (2014)
51. Z. Xia, C. Liu, N.Z. Gong, Q. Li, Y. Cui, D. Song, Characterizing and detecting malicious accounts in privacy-centric mobile social networks: A case study, in *Proceedings of the 25th ACM SIGKDD International Conference on Knowledge Discovery & Data Mining* (ACM, 2019), pp. 2012–2022
52. S. Xiao, M. Farajtabar, X. Ye, J. Yan, L. Song, H. Zha, Wasserstein learning of deep generative point process models, in *Advances in Neural Information Processing Systems*, pp. 3247–3257 (2017)
53. H. Xu, M. Farajtabar, H. Zha, Learning granger causality for Hawkes processes, in *International Conference on Machine Learning*, pp. 1717–1726 (2016)
54. X. Zhang, J. Zhu, Q. Wang, H. Zhao, Identifying influential nodes in complex networks with community structure. Know.-Based Syst. **42**, 74–84 (2013)
55. K. Zhou, H. Zha, L. Song, Learning social infectivity in sparse low-rank networks using multi-dimensional Hawkes processes. Artif. Intell. Stat. 641–649 (2013)
56. K. Zhu, L. Ying, Information source detection in the sir model: A sample-path-based approach. IEEE/ACM Trans. Netw. **24**(1), 408–421 (2016)

Chapter 2
Characterizing Pathogenic Social Media Accounts

2.1 Introduction

Online social media play major role in dissemination of information. However, recent years have witnessed evidence of spreading huge amount of harmful disinformation on social media and manipulating public opinion on the Web, attributed to accounts dedicated to spreading malicious information. These accounts are referred to "Pathogenic Social Media" (PSM) accounts and can pose threats to social media firms and general public. PSMs are usually controlled by terrorist supporters, water armies or fake news writers and they are owned by either real users or bots who seek to promote or degrade certain ideas by utilizing large online communities of supporters to reach their goals. Identifying PSM accounts could have immediate applications, including countering terrorism [13, 14], fake news detection [10, 11], and water armies detection [5, 6].

In order to better understand the behavior and impact of PSM accounts on the Web and normal users, and be able to counter their malicious activity, social media authorities need to deploy certain capabilities which could ultimately lead to reducing their threats. To make this happen, social media platforms are required to design sophisticated techniques that could automatically detect and suspend these accounts as quickly as possible, before they can reach their vast audience and spread malicious content. However, for the most part, the social media firms usually rely on reports they receive from their normal users or even their assigned teams to manually shut down these accounts. First of all, this mechanism is not always feasible due to the limited manpower and since not many real users are willing to put aside time and report the malicious activities. Also, it cannot be done in a timely manner since it takes time for these firms to review the reports and decide whether they are legit or not. On the other hand, the fact that these accounts simply return to social media using different accounts or even migrate to other social media makes all these efforts almost useless. Therefore, the burden falls to automatic approaches that can identify these malicious actors on social media.

© The Author(s), under exclusive license to Springer Nature Switzerland AG 2021
H. Alvari et al., *Identification of Pathogenic Social Media Accounts*, SpringerBriefs
in Computer Science, https://doi.org/10.1007/978-3-030-61431-7_2

This chapter aims to distinguish PSM accounts from their counterparts, i.e., normal users,[1] by analyzing their (1) resharing behavior [17] and (2) their posted URLs [1]. For the former, a probabilistic causal inference framework based on *Suppes' theory of probabilistic causation* [18] is introduced to estimate causality of PSM and normal users and hence distinguish between them [17]. The central concept to this theory is *prima facie causes*: an event to be recognized as a cause must occur before the effect and must lead to an increase of the likelihood of observing the effect. Suppes' theory is chosen over other causality approaches (see [16] for more details) due to its less complex computation requirements. For the latter, a mathematical technique known as "Hawkes processes" [12] is utilized to quantify the impact of PSM users on the greater Web based on their posted URLs on Twitter. Hawkes processes are special forms of point processes and have shown promising results in many problems that require modeling complicated event sequences where historical events have impact on future ones, including financial analysis [4], seismic analysis [7], and social network modeling [21] to name a few.

This study uses an ISIS-related dataset from Twitter used in the previous studies [1–3, 17]. The dataset contains an *action log* of users in the form of cascades of retweets. We also consider URLs posted by two groups of users: (1) PSM accounts and (2) normal users. The URLs can belong to any platform including the major social media (e.g., facebook.com), mainstream news (e.g., nytimes.com), and alternative news outlets (e.g., rt.com). For each group of users, we fit a multi-dimensional Hawkes processes model wherein each process corresponds to a platform referenced in at least one tweet. Furthermore, every process can influence all the others including itself, which allows estimating the strength of connections between each of the social media platforms and news sources, in terms of how likely an event (i.e., the posted URL) can cause subsequent events in each of the groups. In other words, in this chapter we are interested to investigate if a given URL u_1 has influence on another URL u_2 (i.e., $u_1 \rightarrow u_2$) and thus can trigger subsequent events.

This chapter makes the following main observations:

- We introduce a series of causality-based metrics based on *Suppes' theory of probabilistic causation* for distinguishing PSM users from their counterparts. In the subsequent chapters, we will leverage these metrics in supervised, unsupervised, and semi-supervised settings for identification of PSM users.
- We observe that PSM accounts have higher causality values compared to non-PSM users.
- Among all platforms studied here, URLs shared from Facebook and alternative news media contribute the most to the dissemination of malicious information from PSM accounts. Simply put, they had the largest impact on making a message viral and causing the subsequent events.

---

[1] Throughout this book, we may use terms *normal*, *non-PSM*, or *regular users* interchangeably to refer to the accounts that do not intend to do harm to the public and social media.

- Posts that are tweeted by the PSM accounts and contained URLs from Facebook demonstrate more influence on the subsequent retweets containing URLs from YouTube, in contrary to the other way around. This means that ultimately tweets with URLs from Facebook will highly likely end up inducing more external impulse on YouTube than YouTube might have on Facebook.
- URLs posted by the normal users have nearly the same impact on the subsequent events regardless of the social media or news outlet used. This basically means that normal users do not often prefer specific social media or news sources over the others.

2.2 Frameworks

Here, we set out to understand PSM and non-PSM users by investigating their (1) resharing behavior and (2) the impact of their posted URLs on the greater Web. We first study the former by leveraging probabilistic causal inference and then investigate the latter by utilizing Hawkes processes.

2.2.1 Probabilistic Causal Inference

Throughout this book we shall represent cascades as an "action log" ($Actions$) of tuples where each tuple $(u, m, t) \in Actions$ corresponds with a user $u \in U$ posting message $m \in M$ at time $t \in T$, following the convention of [9]. We assume that set M includes posts/repost of a certain original tweet or message. For a given message, we only consider the first occurrence of each user. We define $Actions_m$ as a subset of $Actions$ for a specific message m. Formally, we define it as $Actions_m = \{(u', m', t') \in Actions \text{ s.t. } m' = m\}$ [17].

Definition 2.1 (m-Participant) For a given $m \in M$, user u is an m-**participant** if there exists t such that $(u, m, t) \in Actions$.

Note that the users posting tweet/retweet in the early stage of cascades are the most important ones since they play a significant role in advertising the message and making it viral. For a given $m \in M$, we say m-participant i "precedes" m-participant j if there exists $t < t'$ where $(i, m, t), (j, m, t') \in Actions$. Thus, we define *key users* as a set of users adopting a message in the early stage of its life span. We formally define *key user* as follows:

Definition 2.2 (Key User) For a given message m, m-participant i, and $Actions_m$, we say user i is a **key user** iff user i precedes at least ϕ fraction of m-participants (formally: $|Actions_m| \times \phi \leq |\{j | \exists t' : (j, m, t') \in Actions_m \wedge t' > t\}|$, $(i, m, t) \in Actions_m$), where $\phi \in (0, 1)$.

The notation $|\cdot|$ denotes the cardinality of a set. All messages are not equally important. That is, only a small portion of them gets popular. We define *viral messages* as follows:

Definition 2.3 (Viral Messages) For a given threshold θ, we say that a message $m \in M$ is **viral** iff $|Actions_m| \geq \theta$. We use M_{vir} to denote the set of viral messages.

The Definition 2.3 allows us to compute the prior probability of a message (cascade) going viral as follows:

$$\rho = \frac{|M_{vir}|}{|M|} \tag{2.1}$$

We also define the probability of a cascade m going viral given some user i was involved as

$$p_{m|i} = \frac{|\{m \in M_{vir} \; s.t. \; i \; is \; a \; key \; user\}|}{|\{m \in M \; s.t. \; i \; is \; a \; key \; user\}|} \tag{2.2}$$

We are also concerned with two other measures. First, the probability that two users i and j tweet or retweet viral post m chronologically, and both are key users. In other words, these two users are making post m viral.

$$p_{i,j} = \frac{|\{m \in M_{vir} | \exists t, t' \; where \; t < t' \; and \\ (i, m, t), (j, m, t') \in Actions\}|}{|m \in M | \exists t, t' \; where \; (i, m, t), (j, m, t') \in Actions|} \tag{2.3}$$

Second, the probability that key user j tweets/retweets viral post m and user i does not tweet/retweet earlier than j. In other words, only user j is making post m viral.

$$p_{\neg i,j} = \frac{|\{m \in M_{vir} | \exists t' \; s.t. \; (j, m, t') \in Actions \; and \\ \not\exists t \; where \; t < t', \; (i, m, t) \in Actions\}|}{|\{m \in M | \exists t' \; s.t. \; (j, m, t') \in Actions \; and \\ \not\exists t \; where \; t < t', \; (i, m, t) \in Actions\}|} \tag{2.4}$$

Knowing the action log, we aim to find a set of pathogenic social media (PSM) accounts. These users are associated with the early stages of large information cascades and, once detected, are often deactivated by a social media firm. In the causal framework, we introduce a series of causality-based metrics for identifying PSM users.

We adopt the causal inference framework previously introduced in [15, 18]. We expand upon that work in two ways: (1) we adopt it to the problem of identifying PSM accounts and (2) we extend their single causal metric to a set of metrics. Multiple causality measurements provide a stronger determination of significant causality relationships. For a given viral cascade, we seek to identify potential users who likely *cause* the cascade viral. We first require an initial set of criteria for such a causal user. We do this by instantiating the notion of Prima Facie causes to our particular use case below:

Definition 2.4 (Prima Facie Causal User) A user u is a prima facie causal user of cascade m iff: User u is a key user of m, $m \in M_{vir}$, and $p_{m|u} > \rho$.

For a given cascade m, we will often use the language *prima facie causal user* to describe user i is a prima facie cause for m to be viral. In determining if a given prima facie causal user is causal, we must consider other "related" users. In this work, we say i and j are m-related if (1.) i and j are both prima facie causal users for m, (2.) i and j are both key users for m, and (3.) i precedes j. Hence, we will define the set of "related users" for user i (denoted $R(i)$) as follows:

$$R(i) = \{j \ s.t. \ j \neq i \ , \exists m \in M \ s.t. \ i, j \ are \ m - related\} \qquad (2.5)$$

Therefore, $p_{i,j}$ in (2.3) is the probability that cascade m goes viral given both users i and j, and $p_{\neg i,j}$ in (2.4) is the probability that cascade m goes viral given key user j tweets/retweets it while key user i does not tweet/retweet m or precedes j. The idea is that if $p_{i,j} - p_{\neg i,j} > 0$, then user i is more likely a cause than j for m to become viral. We measure *Kleinberg–Mishra causality* ($\epsilon_{K\&M}$) as the average of this quantity to determine how causal a given user i is as follows:

$$\epsilon_{K\&M}(i) = \frac{\sum_{j \in R(i)} (p_{i,j} - p_{\neg i,j})}{|R(i)|} \qquad (2.6)$$

Intuitively, $\epsilon_{K\&M}$ measures the degree of causality exhibited by user i. Additionally, we find it useful to include a few other measures. We introduce *relative likelihood causality* (ϵ_{rel}) as follows:

$$\epsilon_{rel}(i) = \frac{\sum_{j \in R(i)} S(i, j)}{|R(i)|} \qquad (2.7)$$

$$S(i, j) = \begin{cases} (\frac{p_{i,j}}{p_{\neg i,j} + \alpha}) - 1, & p_{i,j} > p_{\neg i,j} \\ 0, & p_{i,j} = p_{\neg i,j'} \\ 1 - (\frac{p_{\neg i,j}}{p_{i,j}}), & \text{otherwise} \end{cases} \qquad (2.8)$$

where α is infinitesimal. Relative likelihood causality metric assesses the relative difference between $p_{i,j}$ and $p_{\neg i,j}$. This helps us to find new users that may not be prioritized by $\epsilon_{K\&M}$. We also find that if a user is mostly appearing after those with the high value of $\epsilon_{K\&M}$, then it is likely to be a PSM account. One can consider all possible combinations of events to capture this situation. However, this approach is computationally expensive. Therefore, we define $\mathscr{Q}(j)$ as follows:

Table 2.1 Related users $R(.)$ (2.5) of cascades $\tau_1 = \{A, B, C, D, E, F, G, H\}$ and $\tau_2 = \{N, M, C, A, H, V, S, T\}$

User	R_{τ_1}	R_{τ_2}	R
A	$\{B, C, D, E, F\}$	$\{H, V\}$	$\{B, C, D, E, F, H, V\}$
B	$\{C, D, E, F\}$	$\{\}$	$\{C, D, E, F\}$
C	$\{D, E, F\}$	$\{A, H, V\}$	$\{A, D, E, F, H, V\}$
D	$\{E, F\}$	$\{\}$	$\{E, F\}$
E	$\{F\}$	$\{\}$	$\{F\}$
N	$\{\}$	$\{M, C, A, H, V\}$	$\{A, C, H, M, V\}$
M	$\{\}$	$\{C, A, H, V\}$	$\{A, C, H, V\}$
H	$\{\}$	$\{V\}$	$\{V\}$

Table 2.2 Set $\mathscr{Q}(.)$ of users
Table 2.1 in (2.9)

User	Total
A	$\{C, N, M\}$
B	$\{A\}$
C	$\{A, B, N, M\}$
D	$\{A, B, C\}$
E	$\{A, B, C, D\}$
N	$\{\}$
M	$\{N\}$
H	$\{A, C, N, M\}$

$$\mathscr{Q}(j) = \{i \ s.t. \ j \in R(i)\} \tag{2.9}$$

Consider the following example:

Example 1 Consider two cascades (actions) $\tau_1 = \{A, B, C, D, E, F, G, H\}$ and $\tau_2 = \{N, M, C, A, H, V, S, T\}$ where the capital letters signify users. We aim to relate key users while $\phi = 0.5$ (Definition 2.2). Table 2.1 shows the related users $R(.)$ for each cascade. Note that the final set $R(.)$ for each user is the union of all sets from the cascades. Set $Q(.)$ for the users of Table 2.1 are presented in Table 2.2.

Accordingly, we define *neighborhood-based causality* (ϵ_{nb}) as the average $\epsilon_{K\&M}(i)$ for all $i \in Q(j)$ as follows:

$$\epsilon_{nb}(j) = \frac{\sum_{i \in \mathscr{Q}(j)} \epsilon_{K\&M}(i)}{|\mathscr{Q}(j)|} \tag{2.10}$$

The intuition behind this metric is that accounts who are retweeting a message that was tweeted/retweeted by several causal users are potential for PSM accounts. We also define the *weighted neighborhood-based causality* (ϵ_{wnb}) as follows:

$$\epsilon_{wnb}(j) = \frac{\sum_{i \in \mathscr{Q}(j)} w_i \times \epsilon_{K\&M}(i)}{\sum_{i \in \mathscr{Q}(j)} w_i} \tag{2.11}$$

The intuition behind the metric ϵ_{wnb} is that the users in $\mathcal{Q}$ may not have the same impact on user j and thus different weights w_i are assigned to each user i with $\epsilon_{K\&M}(i)$.

2.2.1.1 ISIS-A Dataset

Our dataset consists of ISIS-related tweets/retweets in Arabic gathered from Feb. 2016 to May 2016, hereafter called ISIS-A dataset. This dataset includes tweets and the associated information such as user ID, retweet ID, hashtags, content, date, and time. About 53M tweets are collected based on the 290 hashtags such as Terrorism, State of the Islamic-Caliphate, Rebels, Burqa State, and Bashar-Assad, Ahrar Al-Sham, and Syrian Army. In this work, we only use tweets (more than 9M) associated with viral cascades (Fig. 2.1). The statistics of the dataset are presented in Table 2.3 discussed in details below.

Cascades In this work, we aim to identify PSM accounts—which in this dataset are mainly social bots or terrorism-supporting accounts that participate in viral cascades. The tweets that have been retweeted from 102 to 18,892 times. This leads to more than 35k cascades which are tweeted or retweeted by more than 1M users. The distribution of the number of cascades vs cascade size is illustrated in Fig. 2.2. There are users that retweet their own tweet or retweet a post several times, we only

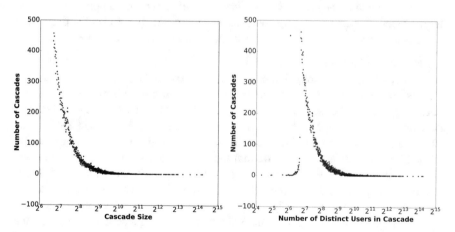

Fig. 2.1 Distribution of cascades vs cascade size for (left) all users and (right) distinct users

Table 2.3 Statistics of the dataset

Name	Values
Tweets	9,092,978
Cascades	35,251
Users	1,249,293
Generator users	8,056

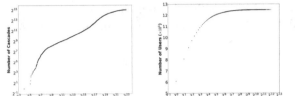

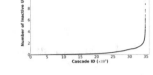

Fig. 2.2 Left to right: Cumulative distribution of duration of cascades. Cumulative distribution of user's occurrence in the dataset. Total inactive users in every cascade

Table 2.4 Status of a subset of the users in the dataset

Name	Active	Inactive	Total
Users	723,727	93,770	817,497
Generator users	7,243	813	8,056

consider the first tweet/retweet of each user for a given cascade. In other words, duplicate users are removed from the cascades, which make the size of the viral cascades from 20 to 18,789 as shown in Fig. 2.2. The distribution of the cascades over the cascade life span is illustrated in Fig. 2.2. Cascades took from 16 seconds to more than 94 days to complete.

Users There are more than 1M users that have participated in the viral cascades. Figure 2.2 demonstrates the cumulative distribution of the number of times a user have participated in the cascades. As it is shown, the larger the support value is, the less number of users exists. Moreover, users have tweeted or retweeted posts from 1 to 3,904 times and on average each user has participated more than 7 times.

User's Current Status We select *key users* that have tweeted or retweeted a post in its early life span—among first half of the users (according to Definition 2.2, $\phi = 0.5$). For labeling, we check through Twitter API to examine whether the users have been suspended (labeled as PSM) or they are still active (labeled as normal) [19]. Accounts are not active if they are suspended or deleted. More than 88% of the users are active as shown in Table 2.4. The statistics of the generator users are also reported. Generator users are those that have initiated a viral cascade. As shown, 90% of the generator users are active as well. Moreover, there are a significant number of cascades with hundreds of inactive users. The number of inactive users in every cascade is illustrated in Fig. 2.2. Inactive users are representative of automatic and terrorism accounts aiming to disseminate their propaganda and manipulate the statistics of the hashtags of their interest.

Generator Users In this part, we only consider users that have generated (started) the viral tweets. According to Table 2.4, there are more than 7k active and 800 inactive generator users. That is, more than 10% of the generator users are suspended or deleted, which means they are potentially automated accounts. The distribution of the number of tweets generated by generator users shows that most of the users (no matter active and inactive) have generated a few posts (less than or equal to 3), while only a limited number of users are with a large number of tweets.

2.2.1.2 Causality Analysis

Here we examine the behavior of the causality metrics. We analyze users considering their current account status in Twitter. We label a user as active (inactive) if the account is still active (suspended or deleted).

Kleinberg–Mishra Causality We study the users that get their causality value of $\epsilon_{K\&M}$ greater than or equal to 0.5. As expected, inactive users exhibit different distribution from active users (Fig. 2.3 (Top Left)). We note that significant differences are present—more than 75% of the active users are distributed between 0.5 and 0.62, while more than 50% of the inactive users are distributed from 0.75 to 1. Also, inactive users have larger values of mean and median than active ones. Note that number of active and inactive users are 404,536 and 52,452. This confirms that this metric is a good indicator to discriminate PSM users from the normal users.

Relative Likelihood Causality This metric magnifies the interval between every pairs of the probabilities that measures the causality of the users; therefore, the values vary in a wide range. Figure 2.3 (Top Right) displays the distribution of users having relative likelihood causality of greater than or equal to two. In this metric, 1,274 users get very large values. For the sake of readability, very large values are replaced with 34.0. More than 50% of the inactive users get values greater than 32, while the median of active users is 2.48. More than 75% of the active users are distributed in the range of $(2, 4)$. Note that number of active and inactive users in this figure are 3,563 and 1,041, respectively. That is, using this metric and filtering users with the relative likelihood greater than a threshold leads to the good precision. For example, the threshold is set to 2—the precision is more than 0.22 for inactive class. Considering users with a very large value leads to the precision of more than 0.5 and uncovering a significant number of PSMs—638 inactive users.

Neighborhood-Based Causality We study the users that get their causality value of ϵ_{nb} greater than or equal to 0.5. As expected, inactive users exhibit different distribution from active users as shown in Fig. 2.3 (Bottom Left). Also, inactive users are mostly distributed in the higher range and have larger values of mean and median than active ones. More than 75% of the active users are distributed between 0.5 and 0.6, while more than 50% of the inactive users are distributed from 0.6 to 1. Therefore, increasing the threshold results in the higher precision for the PSM users. Note that the number of active and inactive users are 85,864 and 10,165.

Weighted Neighborhood-Based Causality This metric is the weighted version of the previous metric (ϵ_{nb}). We assign weight to each user in proportion to her participation rate in the viral cascades. Figure 2.3 (Bottom Right) shows the distribution of users with ϵ_{wnb} greater than or equal to 0.5. This metric also displays

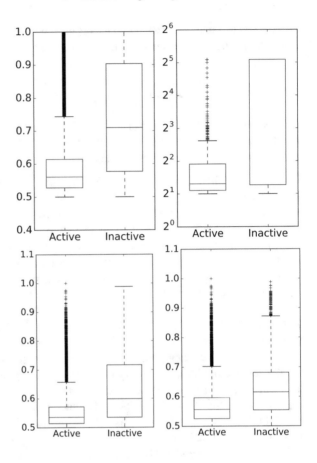

Fig. 2.3 Distribution of various causality metrics for active and inactive users: $\epsilon_{K\&M} \geq 0.5$ (Top left), $\epsilon_{rel} \geq 2$ (Top right), $\epsilon_{nb} \geq 0.5$ (Bottom left), $\epsilon_{wnb} \geq 0.5$ (Bottom right)

different distribution for active and inactive users. More than 75% of the active users are distributed between 0.5 and 0.6, while more than 50% of the inactive users are distributed from 0.6 to 1. Note that the number of active and inactive users of ϵ_{wnb} are 52,346 and 16,412. In other words, this metric achieves the largest precision compared to other metrics, 0.24. Clearly, increasing the threshold results in the higher precision for the PSMs.

2.2.2 Hawkes Processes

In many scenarios, one needs to deal with timestamped events such as the activity of users on a social network recorded in continuous time. An important task then is to estimate the influence of the nodes based on their timestamp patterns [8]. Point process is a principled framework for modeling such event data, where the dynamic of the point process can be captured by its conditional intensity function as follows:

$$\lambda(t) = \lim_{\Delta t \to 0} \frac{\mathbb{E}(N(t + \Delta t) - N(t)|\mathcal{H}_t)}{\Delta t} = \frac{\mathbb{E}(dN(t)|\mathcal{H}_t)}{dt} \tag{2.12}$$

where $\mathbb{E}(dN(t)|\mathcal{H}_t)$ is the expectation of the number of events happened in the interval $(t, t+dt]$ given the historical observations $\mathcal{H}_t$ and $N(t)$ records the number of events before time t. Point process can be equivalently represented as a counting process $N = \{N(t)|t \in [0, T]\}$ over the time interval $[0, T]$.

The Hawkes process framework [12] has been used in many problems that require modeling complicated event sequences where historical events have impact on future ones. Examples include but are not limited to financial analysis [4], seismic analysis [7], and social network modeling [21]. One-dimensional Hawkes process is a point process N_t with the following particular form of intensity function:

$$\lambda(t) = \mu + a \int_{-\infty}^{t} g(t - s)dN_s = \mu + a \sum_{i:t_i < t} g(t - t_i) \tag{2.13}$$

where $\mu > 0$ is the exogenous base intensity (i.e., background rate of events), t_i are the time of events in the point process before time t, and $g(t)$ is the decay kernel.

In this work, we use exponential kernel of the form $g(t) = we^{-wt}$, but adapting to the other positive forms is straightforward. The second part of the above formulation captures the self-exciting nature of the point processes—the occurrence of events in the past has a positive impact on the future ones. Given a sequence of events $\{t_i\}_{i=1}^{n}$ observed in $[0, T]$ and generated from the above intensity function, the log-likelihood function can be obtained as follows [21]:

$$\mathcal{L} = \log \frac{\prod_{i=1}^{n} \lambda(t_i)}{\exp \int_0^T \lambda(t)dt} = \sum_{i=1}^{n} \log \lambda(t_i) - \int_0^T \lambda(t)dt \tag{2.14}$$

In this work, we focus on multi-dimensional Hawkes processes which is defined by a U-dimensional point process $N_t^u, u = 1, \ldots, U$. In other words, we have U Hawkes processes coupled with each other—each Hawkes process corresponds to one of the platforms and the influence between them is modeled using the mutually-exciting property of the multi-dimensional Hawkes processes. We formally define the following formulation to model the influence of different events on each other:

$$\lambda_u(t) = \mu_u + \sum_{i:t_i < t} a_{uu_i} g(t - t_i) \tag{2.15}$$

where $\mu_u \geq 0$ is the base intensity for the u-th Hawkes process. The coefficient $a_{uu_i} \geq 0$ captures the mutually-exciting property between the u-th and u_i-th processes. Larger value of a_{uu_i} shows that events in the u_i-th dimension are more likely to trigger an event in u-th dimension in future. More intuitively, an event on one point process can cause an impulse response on other processes, which increases the probability of an event occurring above the processes' background rates. We

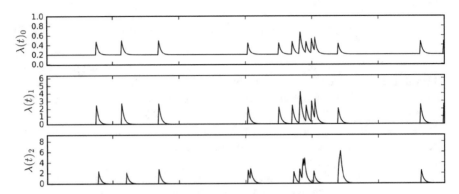

Fig. 2.4 Illustration of the Hawkes Process. Events induce impulse on other processes and cause child events. Background event in e_0 induces impulse on responses on processes e_1 and e_2

reiterate that in this study each URL is attributed to an event, i.e., if the URL u_1 triggers the URL u_2 (i.e., $u_1 \rightarrow u_2$), then $a_{u_2 u_1} \geq 0$

In Fig. 2.4, we depict a multivariate example of three different streams of events, e_0, e_1, and e_2. As illustrated, e_0 is caused by the background rate $\lambda(t)_0$ and has an influence on itself and e_1. On the other hand, e_1 is caused by $\lambda(t)_1$ and has an influence on e_2. Simply put, a background event in e_0 induces impulse on responses on processes e_1 and e_2. Accordingly, the caused child event in e_1 leads to another child event in e_2.

We consider an infectivity matrix $A = [a_{u u_i}] \in \mathbb{R}^{U \times U}$ which collects the self-triggering coefficients between Hawkes processes, and $U = 7$ is the number of processes (i.e., platforms) in our work. Each entry in this matrix indicates the strength of influence each platform has on other platforms. Our ultimate goal in this work is to estimate the infectivity matrix as it reflects the estimated influence of each platform on others. Next, we will provide the methodology that we follow to estimate the influence of the URLs on each other.

We aim to assess the influence of the PSM accounts in our dataset via their posted URLs. We consider the URLs posted by two groups of users: (1) PSM accounts and (2) normal users. For both groups, we fit a Hawkes model with $K = 7$ point processes each for the seven categories of social media and news outlets discussed earlier. In each of the Hawkes models, every process is able to influence all the others including itself, which allows us to estimate the strength of connections between each of the seven categories for both groups of users, in terms of how likely an event (i.e., the posted URL) can cause subsequent events in each of the groups.

We use the ADM4 algorithm presented by [21] and follow the methodology presented by [20] for fitting the Hawkes processes for both PSM and normal users. ADM4 [21] is an efficient optimization that estimates the parameters A and μ by maximizing the regularized log-likelihood $\mathscr{L}(A, \mu)$:

$$\min_{A \geq 0, \mu \geq 0} -\mathscr{L}(A, \mu) + \lambda_1 ||A||_* + \lambda_2 ||A||_1 \tag{2.16}$$

where $\mathscr{L}(A, \mu)$ can be obtained by substituting $\lambda_u(t)$ from Eq. 2.15 into Eq. 2.14. Also, $||A||_*$ is the nuclear norm of matrix A, and is defined as the sum of its singular value.

We consider two different sets of URLs posted by the PSM accounts and normal users by selecting URLs that have at least one event in Twitter (i.e., posted by a user). For each group, we construct a matrix $W \in \mathbb{N}^{T \times U}$ with $U = 7$, whose entries are sequences of events (i.e., posted URLs) observed during a time period T. We note that each sequence of events is of the form $\mathscr{S} = \{(t_i, u_i)\}_{i=1}^{n_i}$ where n_i is the number of the events occurring at the u_i-th dimension (i.e., URLs posted containing one of the 7 platforms).

2.2.2.1 ISIS-B Dataset

Here we use a subset of the previous dataset with 2.8M ISIS-related tweets/retweets in Arabic, hereafter called ISIS-B dataset. In this dataset, about 600K tweets have at least one URL (i.e., event) referencing one of the social media platforms or news outlets. There are about 1.4M of paired URLs which we denote by $u_1 \rightarrow u_2$ and indicate a retweet (with the URL u_2) of the original tweet (with the URL u_1).

In this study, we are interested in investigating the impact of the URL u_1 on u_2. Accordingly, the dataset contains 35K cascades (i.e., sequences of events) of different sizes and durations, some of which contain paired URLs in the aforementioned form. After pre-processing and removing duplicate users from cascades (those who retweet themselves multiple times), cascades sizes (i.e., number of associated postings) vary between 20 to 9,571 and take from 10 seconds to 95 days to finish. The log-log distribution of cascades vs. cascade size and the cumulative distribution of duration of cascades are depicted in Fig. 2.5.

The statistics of the dataset are presented in Table 2.5. As before, we check through Twitter API to examine whether the users have been suspended (labeled as PSM) or are still active (labeled as normal) [19]. According to Table 2.5, 11% of the users in our dataset are PSMs and others are normal. We also depict the total number of PSM accounts that have been suspended by Twitter in each cascade, in Fig. 2.6. Finally, to reiterate, we note that these accounts mostly get suspended manually by Twitter based on reports the platform receives from its own users.[2]

---

[2]https://blog.twitter.com/official/en_us/a/2016/an-update-on-our-efforts-to-combat-violent-extremism.html.

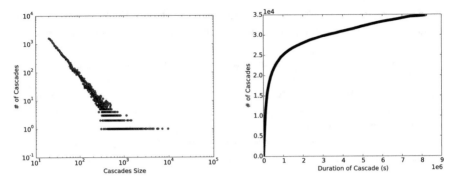

Fig. 2.5 (Left) Log-log distribution of cascades vs. cascade size. (Right) Cumulative distribution of duration of cascades

Table 2.5 Description of the dataset

Name	Value	
# of Cascades	35K	
# of Tweets/Retweets	2.8M	
	PSM	Normal
# of Users	64,484	536,609
# of Single URLs	104,948	536,046
# of Paired URLs	200,892	1,123,434

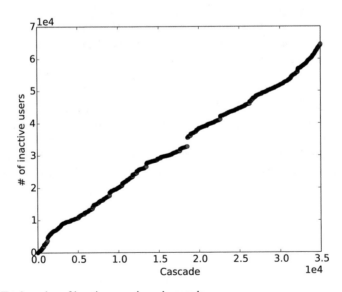

Fig. 2.6 Total number of inactive users in each cascade

Social Media Platforms and News Outlets

Twitter deploys a URL shortener technique to leave more space for content and protect users from malicious sites.[3] To obtain the original URLs, we use a URL unshortening tool[4] to obtain the original links contained in the tweets in our dataset.

We consider a number of major and well-known social media platforms including Twitter, Facebook, Instagram, Google, and YouTube. About the dichotomy of mainstream and alternative media, it is notable to mention that most criteria for determining whether a news source counts as either of them are based on a number of factors including but not limited to the content and whether or not it is corporate owned.[5] However, a key difference between these two sources of media comes from the fact that all of mainstream media are profit-oriented, in contrast to the alternative media. We further note that for the most part, mainstream media is considered as a more credible source than alternative media, although the reputation has been recently tainted by the fake news.

In this work, following the commonsense, we consider popular news outlets such as The New York Times, and The Wall Street Journal as mainstream and less popular ones as alternatives. In Table 2.6, we summarize the total number of paired URLs (i.e., $u_1 \rightarrow u_2$) in which the original URL (i.e., u_1) corresponds to each social media platform with at least one event in our dataset. We also summarize in Table 2.7, the total number of paired URLs whose original URL belongs to the mainstream and alternative news sources. In Table 2.8, we see the breakdown of number of paired URLs for the PSM and normal users. We further demonstrate in Table 2.9 some examples of the mainstream and alternative news URLs occurrence used in this work.

Table 2.6 Social media platform's total number of paired URLs of the form $u_1 \rightarrow u_2$ with at least one event in the dataset for the PSM and normal users

Platform	PSM	Normal
Twitter	139,940	918,803
Facebook	878	4,017
Instagram	0	2,857
Google	163	132
YouTube	24,724	72,890

Table 2.7 News sources' total paired URLs ($u_1 \rightarrow u_2$) with at least one event in the dataset for the PSM and normal users

News source	PSM	Normal
Mainstream	0	286
Alternatives	35,187	124,449

---

[3] https://help.twitter.com/en/using-twitter/url-shortener.

[4] https://github.com/skevas/unshorten.

[5] https://smallbusiness.chron.com/mainstream-vs-alternative-media-21113.html.

Table 2.8 Total number of paired URLs of the form $u_1 \rightarrow u_2$ with at least one event for PSM/normal users and for all platforms

	$\rightarrow$ Twitter	$\rightarrow$ Facebook	$\rightarrow$ Instagram	$\rightarrow$ Google	$\rightarrow$ YouTube	$\rightarrow$ Mainstream	$\rightarrow$ Alternatives
Twitter $\rightarrow$	109,354/766,617	598/3,843	229/2,461	120/382	11,992/59,889	90/688	17,557/84,923
Facebook $\rightarrow$	655/3,108	4/41	3/9	2/1	87/281	0/1	127/576
Instagram $\rightarrow$	0/2,362	0/11	0/25	0/2	0/161	0/2	0/294
Google $\rightarrow$	134/74	0/0	0/1	0/0	12/53	0/0	17/4
YouTube $\rightarrow$	14,004/56,545	132/312	23/211	22/32	6,799/7,529	13/48	3,731/8,213
Mainstream $\rightarrow$	0/189	0/1	0/1	0/0	0/13	0/1	0/81
Alternatives $\rightarrow$	21,047/95,641	145/767	45/318	59/64	3,862/9,199	26/122	10,003/18,338

Table 2.9 Examples of mainstream and alternative news

Mainstream	Alternatives
https://www.nytimes.com	https://www.rt.com
https://www.reuters.com	https://www.arabi21.com
https://www.wsj.com	https://www.7adramout.net
https://www.nbcnews.com	https://www.addiyar.com
https://www.ft.com	https://zamnpress.com

2.2.2.2 Experimental Results

Here, we conduct experiments to gauge the effectiveness of Hawkes process for modeling influence of PSMs.

Settings

In this work, we adopt the ADM4 algorithm [21] which implements parametric inference for Hawkes processes with an exponential kernel and a mix of Lasso and nuclear regularization. We initialize infectivity matrix A, base intensities μ, and decays $\beta \in \mathbb{R}$ randomly.

We further set the number of nodes $U = 7$ to reflect the 7 platforms used in this study. Level of penalization is set to $C = 1000$, and the ratio of Lasso-Nuclear regularization mixing parameter is set to 0.5. Finally, maximum number of iterations for solving the optimization is set to 50 and the tolerance of solving algorithm is set to $1e - 5$.

Temporal Analysis

Here, we present the differences between the PSM accounts in our dataset with their counterparts, normal users through temporal analysis of their posted URLs.

In Fig. 2.7, we depict the daily occurrence of the paired URLs over the span of 43 days for both PSM and normal users. Recall from the previous section that our dataset has a larger number of normal users and higher number of the posted URLs compared to the PSM accounts. Therefore, it is reasonable to observe more activity from normal users than PSMs. For both groups of users, we observe a similar trend in occurrence of spikes and their durations. As it is seen, distinguishing between PSMs and normal users merely based on the occurrence of URLs and their patterns is not reliable. Therefore, we set out to conduct experiments using a more sophisticated statistical technique known as "Hawkes Process" in the next section.

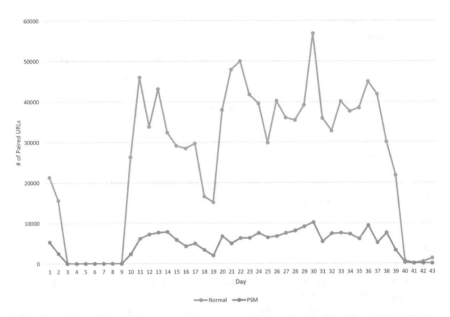

Fig. 2.7 Number of paired URLs posted by the PSM and normal users in our dataset. Note that number of normal users in our dataset is higher than the PSM accounts

Results

We estimate infectivity matrix for both PSM and normal users by fitting the Hawkes model described earlier. In our study, this matrix characterizes the strength of the connections between the platforms and news sources. More specifically, each weight value represents the connection strength from one platform to another. In other words, each entry in this matrix can be interpreted as the expected number of subsequent events that will occur on the second group after each event on the first [20]. In Fig. 2.8, we depict the estimated weights for all paired URLs for both PSM and normal users. Looking at the weights in both of the plots, we realize that greater weights belong to processes that have impact on Twitter, i.e., "$\rightarrow Twitter$". This implies that both of the groups in our Twitter dataset often post URLs that ultimately have greater impact on Twitter.

Overall, we observe the followings:

- URLs referencing all platforms and posted by the PSMs and regular users, mostly trigger URLs that contain the Twitter domain.
- Among all platforms studied here, URLs shared from facebook.com and alternative news media contributed the most to the dissemination of malicious information from PSM accounts. In other words, they had largest impact on making a message viral and causing the subsequent events.
- Posts that were tweeted by the PSM accounts and contained URLs from facebook.com demonstrated more influence on the subsequent retweets containing

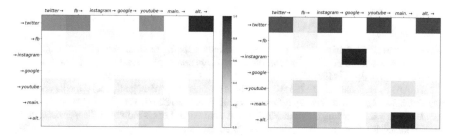

Fig. 2.8 From Left to Right: Estimated infectivity matrices for all paired URLs for PSMs and normal users. Among all URLs, those shared from https://www.Facebook.com and alternative news media had the largest impact on dissemination of malicious messages

URLs from youtube.com, in contrary to the other way around. This means that ultimately tweets with URLs from Facebook will likely end up inducing external impulse on youtube.com. In contrast, URLs posted by the normal users have nearly the same impact on the subsequent events regardless of the social media or news outlet used.

The above mentioned observations demonstrate the effectiveness of leveraging Hawkes process to quantify the impact of URLs posted by PSMs and regular users on the dissemination of content on Twitter. The observations we make here show that PSM accounts and regular users behave differently in terms of the URLs they post on Twitter, in that they have different tastes while disseminating URL links. Accordingly their impact on the subsequent events significantly differ from each other.

2.3   Conclusion

This chapter presented causal inference and Hawkes processes frameworks to differentiate between PSM and normal users. Specifically, first part of the chapter presented a suit of causality-based metrics built upon *Suppes' theory of probabilistic causation* to distinguish between PSM and non-PSM accounts. Our analyses on a real-world ISIS-related dataset from Twitter indicated that in general, PSM accounts have higher causality values compared to regular users.

In the second part, we presented an analysis on a subset of the previous dataset to demonstrate how PSM and normal users usually post on Twitter in terms of the URLs they post. We leveraged a statistical technique known as Hawkes Process for modeling the influence of PSM accounts on dissemination of malicious content on the Web. In this work, we used URLs posted by two groups of users, PSMs and normal users, on major social media and mainstream and alternative news outlets. Overall, our findings indicate that the URLs posted by the PSM accounts have the largest impact if contained either facebook.com or alternative news media. In contrast, their counterparts, i.e., normal users, often post URLs that have nearly the same impact on the Web, no matter what social media or news outlet they use.

References

1. H. Alvari, P. Shakarian, Hawkes process for understanding the influence of pathogenic social media accounts, in *2019 2nd International Conference on Data Intelligence and Security (ICDIS)*, pp. 36–42 (June 2019)
2. H. Alvari, E. Shaabani, P. Shakarian, Early identification of pathogenic social media accounts. *IEEE Intelligent and Security Informatics* (2018). arXiv:1809.09331
3. H. Alvari, E. Shaabani, S. Sarkar, G. Beigi, P. Shakarian, Less is more: Semi-supervised causal inference for detecting pathogenic users in social media, in *Companion Proceedings of The 2019 World Wide Web Conference*, WWW '19 (Association for Computing Machinery, New York, NY, USA, 2019), pp. 154–161
4. E. Bacry, T. Jaisson, J.-F. Muzy, Estimation of slowly decreasing Hawkes kernels: application to high-frequency order book dynamics. Quantitative Finance **16**(8), 1179–1201 (2016)
5. C. Chen, K. Wu, S. Venkatesh, X. Zhang, Battling the internet water army: Detection of hidden paid posters. CoRR, abs/1111.4297 (2011)
6. C. Chen, K. Wu, S. Venkatesh, R.K. Bharadwaj, The best answers? think twice: online detection of commercial campaigns in the CQA forums, in *ASONAM* (2013)
7. D.J. Daley, D. Vere-Jones, *An Introduction to the Theory of Point Processes: Volume II: General Theory and Structure* (Springer Science & Business Media, 2007)
8. M. Gomez-Rodriguez, J. Leskovec, B. Schölkopf, Modeling information propagation with survival theory, in *International Conference on Machine Learning*, pp. 666–674 (2013)
9. A. Goyal, F. Bonchi, L.V. Lakshmanan, Learning influence probabilities in social networks, in *WSDM* (2010)
10. A. Gupta, H. Lamba, P. Kumaraguru, $1.00 per rt #bostonmarathon #prayforboston: Analyzing fake content on twitter, in *2013 APWG eCrime Researchers Summit* (2013)
11. A. Gupta, P. Kumaraguru, C. Castillo, P. Meier, *TweetCred: Real-Time Credibility Assessment of Content on Twitter* (Springer International Publishing, 2014)
12. A.G. Hawkes, Spectra of some self-exciting and mutually exciting point processes. Biometrika **58**(1), 83–90 (1971)
13. M. Khader, *Combating Violent Extremism and Radicalization in the Digital Era*. Advances in Religious and Cultural Studies (IGI Global, 2016)
14. J. Klausen, C. Marks, T. Zaman, Finding online extremists in social networks. CoRR, abs/1610.06242 (2016)
15. S. Kleinberg, B. Mishra, The temporal logic of causal structures. CoRR, abs/1205.2634 (2012)
16. J. Pearl, *Causality: Models, Reasoning and Inference*, 2nd edn. (Cambridge University Press, New York, NY, USA, 2009)
17. E. Shaabani, R. Guo, P. Shakarian, Detecting pathogenic social media accounts without content or network structure, in *2018 1st International Conference on Data Intelligence and Security (ICDIS)* (IEEE, 2018), pp. 57–64
18. P. Suppes, A probabilistic theory of causality (1970)
19. K. Thomas, C. Grier, D. Song, V. Paxson, Suspended accounts in retrospect: an analysis of twitter spam, in *Proceedings of the 2011 ACM SIGCOMM Conference on Internet Measurement Conference* (ACM, 2011), pp. 243–258
20. S. Zannettou, T. Caulfield, E. De Cristofaro, N. Kourtelris, I. Leontiadis, M. Sirivianos, G. Stringhini, J. Blackburn, The web centipede: understanding how web communities influence each other through the lens of mainstream and alternative news sources, in *Proceedings of the 2017 Internet Measurement Conference* (ACM, 2017), pp. 405–417
21. K. Zhou, H. Zha, L. Song, Learning social infectivity in sparse low-rank networks using multi-dimensional Hawkes processes, in *Artificial Intelligence and Statistics*, pp. 641–649 (2013)

Chapter 3
Unsupervised Pathogenic Social Media Accounts Detection Without Content or Network Structure

3.1 Introduction

In Chap. 2, we demonstrated that PSM accounts are key users to propagation of malicious information and formation of malicious campaigns. Existing methods to detect PSM users usually rely on content [9], network structure [6], or a combination of both [7, 8, 12]. However, merely reliance on such information may lead to two challenges. First, network structure which is at the core of many techniques [1, 2, 4, 5, 13] is not always available. For example, the Facebook API does not make this information available without permission of the users (which is likely a non-starter for PSM accounts). Second, the use of content often necessitates the training of a new model for a previously unobserved topic. For example, PSM accounts taking part in elections in the U.S. and Europe will likely leverage different types of content.

In this chapter, we introduce two methods based on causal analysis described previously in Chap. 2 to tackle these issues. The main input to our framework is an *activity log* of user's activities along with their timestamp. We aim to find PSM users who have participated in the *viral* information propagated online in the form of *cascades*. As viral cascades are often very rare, users that cause them are treated suspicious. We leverage causal analysis [11] from previous chapter to address the problem at hand. This chapter makes the following contributions [11]:

- We propose a causal-based framework to detect PSM accounts that does not leverage network structure, cascade path information, content, and user's information.
- We introduce an unsupervised label propagation framework that along with our causal metrics provides a precision of 0.75. The framework significantly outperforms random method with the precision of 0.11, the content-based bot detection method with the precision of 0.13, all features method with the precision of 0.16, and Sentimetrix [12] with the precision of 0.11.

© The Author(s), under exclusive license to Springer Nature Switzerland AG 2021
H. Alvari et al., *Identification of Pathogenic Social Media Accounts*, SpringerBriefs
in Computer Science, https://doi.org/10.1007/978-3-030-61431-7_3

- The proposed framework performs well with larger cascades with many PSM users involved and is able to find more suspicious PSM accounts than the baseline methods.

3.2   Technical Approach

3.2.1   Problem Statements

Our goal is to find the potential PSM accounts from the cascades. Assigning a score to each user and applying threshold-based algorithm is one way of selecting users. In the previous chapter, we defined causality metrics where each of them or combination of them can be a strategy for assigning scores. Users with high values for causality metrics are more likely to be PSM accounts—later we demonstrate the relationship between these measurements and the real world by identifying accounts deactivated eventually. Below, before presenting the problem statements, we first recap the causality metrics from the previous chapter.

We measure *Kleinberg–Mishra causality* ($\epsilon_{K\&M}$) as the average of this quantity to determine how causal a given user i is as follows:

$$\epsilon_{K\&M}(i) = \frac{\sum_{j \in R(i)} (p_{i,j} - p_{\neg i,j})}{|R(i)|} \tag{3.1}$$

Intuitively, $\epsilon_{K\&M}$ measures the degree of causality exhibited by user i. Additionally, we find it useful to include a few other measures. We introduce *relative likelihood causality* (ϵ_{rel}) as follows:

$$\epsilon_{rel}(i) = \frac{\sum_{j \in R(i)} S(i, j)}{|R(i)|} \tag{3.2}$$

$$S(i, j) = \begin{cases} (\frac{p_{i,j}}{p_{\neg i,j} + \alpha}) - 1, & p_{i,j} > p_{\neg i,j} \\ 0, & p_{i,j} = p_{\neg i,j'} \\ 1 - (\frac{p_{\neg i,j}}{p_{i,j}}), & \text{otherwise} \end{cases} \tag{3.3}$$

where α is infinitesimal. Relative likelihood causality metric assesses the relative difference between $p_{i,j}$ and $p_{\neg i,j}$. This helps us to find new users that may not be prioritized by $\epsilon_{K\&M}$. We also find that if a user is mostly appearing after those with the high value of $\epsilon_{K\&M}$, then it is likely to be a PSM account. One can consider all possible combinations of events to capture this situation. However, this approach is computationally expensive. Therefore, we define $\mathcal{Q}(j)$ as follows:

$$\mathcal{Q}(j) = \{i \ s.t. \ j \in R(i)\} \tag{3.4}$$

Next, we define the problem that we are interested to solve:

3.1 (Threshold-Based Problem) Given a causality metric ϵ_k where $k \in \{K\&M, rel, nb, wnb\}$, parameter θ, set of users U, we wish to identify set $\{u\ s.t.\ \forall u \in U,\ \epsilon_k(u) \geq \theta\}$.

We find that considering a set of cascades as a hypergraph where users of each cascade are connected to each other can better model the PSM accounts. The intuition is that densely connected users with high values for causality are the most potential PSM accounts. In other words, we are interested in selecting a user if (1.) it has a score higher than a specific threshold or (2.) it has a lower score but occurs in the cascades where high score users occur. Therefore, we define the *label propagation* problem as follows:

3.2 (Label Propagation Problem) Given a causality metric ϵ_k where $k \in \{K\&M, rel, nb, wnb\}$, parameters θ, λ, set of cascades $\mathscr{T} = \{\tau_1, \tau_1, ..., \tau_n\}$, and set of users U, we wish to identify set $\mathscr{S} : \mathscr{S}_1, \mathscr{S}_2, ..., \mathscr{S}_l, ..., \mathscr{S}_{|U|}$ where $\mathscr{S}_l = \{u | \forall \tau \in \mathscr{T}, \forall u \in (\tau \backslash \mathscr{S}_{l-1}), \epsilon_k(u) \geq (H_\tau^l - \lambda)\}$ and $H_\tau^l = \{min(\epsilon_k(u))\ s.t.\ \forall u \in \tau \wedge u \in \bigcup_{l' \in [1,l)} \mathscr{S}_{l'}\}$.

3.3 Algorithms

3.3.1 Algorithm for Threshold-Based Problems

To calculate causality metrics, we use map-reduce programming model. In this approach, we select users with causality value greater than or equal to a specific threshold. We refer to this approach as the *Threshold-Based Selection Approach*.

3.3.2 Label Propagation Algorithms

Label propagation algorithms [3, 10, 14] iteratively propagate labels of a seed set to their neighbors. All nodes or a subset of nodes in the graph are usually used as a seed set. We propose a Label Propagation Algorithm (Algorithm 1) to solve problem 2. We first take users with causality value greater than or equal to a specific threshold (i.e., 0.9) as the seed set. Then, in each iteration, every selected user u can activate user u' if the following two conditions are satisfied: (1.) u and u' have at least one cascade (action) in common and (2.) $\epsilon_k(u') \geq \epsilon_k(u) - \lambda$, $\lambda \in (0, 1)$. Note that, we set a minimum threshold such as 0.7 so that all users are supposed to satisfy it. In this algorithm, inputs are a set of cascades (actions) $\mathscr{T}$, causality metric ϵ_k and two parameters θ, λ in (0, 1). This algorithm is illustrated by a toy example:

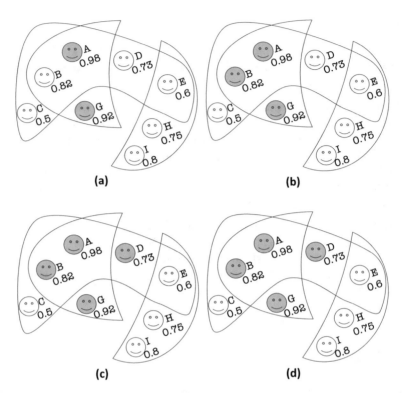

Fig. 3.1 A toy example of Algorithm ProSel. Blue faces depict active users

Example 2 Consider three cascades {{A, B, G}, {A, B, C, D, E, G, H, I}, {E, H, I}} as shown in hypergraph Fig. 3.1. Let us consider the minimum acceptable value as 0.7; in this case, users C and E would not be activated in this algorithm. Assuming two parameters $\theta = 0.9$, $\lambda = 0.1$, both users A and G get activated (Fig. 3.1a). Note that an active user is able to activate inactive ones if (1.) it is connected to the inactive user in the hypergraph, (2.) score of the inactive user meets the threshold. In the next step, only user B will be influenced by G ($0.82 \geq 0.92 - 0.1$) as it is shown in Fig. 3.1b. Then, user D will be influenced by user B ($0.73 \geq 0.82 - 0.1$). In the next step (Fig. 3.1d), the algorithm terminate since no new user is adopted. As it is shown, user I and H are not influenced although they have larger values of ϵ in comparison with user D.

Proposition 3.1 *Given a set of cascades $\mathcal{T}$, a threshold θ, parameter λ, and causality values ϵ_k where $k \in \{K\&M, rel, nb, wnb\}$, ProSel returns a set of users $\mathcal{R} = \{u | \epsilon_k(u) \geq \theta$ or $\exists u'$ s.t. $u', u \in \tau, \epsilon_k(u) \geq \epsilon_k(u') - \lambda$ and u' is picked$\}$. Set $\mathcal{R}$ is equivalent to the set $\mathcal{S}$ in Problem 3.2.*

Proposition 3.2 *The time complexity of Algorithm ProSel is $O(|\mathcal{T}| \times avg(log(|\tau|)) \times |U|)$.*

Algorithm 1 Label propagation algorithm (*ProSel*)

1: **procedure** PROSEL($\mathcal{T}, \epsilon_k, \theta, \lambda$)
2:     $\mathcal{S} = \{(u, \epsilon_k(u)) | \forall u \in U, \epsilon_k(u) \geq \theta\}$
3:     $\mathcal{R} = \mathcal{S}$
4:     $H = \emptyset$
5:     **while** $|\mathcal{S}| > 0$ **do**
6:         $H' = \{(\tau, \epsilon_m) | \forall (\tau, \epsilon) \in H, \; \epsilon_m = min(\epsilon, min(\{\epsilon' = \mathcal{S}_u \; s.t. \; \forall u \in \tau \wedge u \in \mathcal{S}\}))\}$
7:         $H = H' \cup \{(\tau, \epsilon_m) | \forall \tau \in \mathcal{T} \wedge \tau \notin H', \; \epsilon_m = min(\{\epsilon = \mathcal{S}_u \; s.t. \; \forall u \in \tau \wedge u \in \mathcal{S}\})\}$
8:         $\mathcal{S} = \{(u, \epsilon) | \forall \tau \in \mathcal{T}, \; \forall u \in \tau, \; u \notin \mathcal{R}, \; \epsilon_k(u) \geq (H_\tau - \lambda)\}$
9:         $\mathcal{R} = \mathcal{R} \cup \mathcal{S}$
10:    **end while**
11:    **return** $\mathcal{R}$
12: **end procedure**

3.4 Results and Discussion

We implement our code in Scala Spark and Python 2.7x and run it on a machine equipped with an Intel Xeon CPU (1.6 GHz) with 128 GB of RAM running Windows 7. We set the parameter ϕ to label key users 0.5 (Definition 2.2). Thus, we are looking for the users that participate in the action before the number of participants gets twice.

In the following sections, first we look at the existing methods. Then we look at two proposed approaches (see Sect. 3.3): (1) *Threshold-Based Selection Approach*—selecting users based on a specific threshold, (2) *Label Propagation Selection Approach*—selecting by applying Algorithm 1. The intuition behind this approach is to select a user if it has a score higher than a threshold or has a lower score but occurs in the cascades that high score users exist. We evaluate methods based on true positive (TP), false positive (FP), precision, the average (Avg CS), and median (Med CS) of cascade size of the detected PSM accounts. Note that in our problem, precision is the most important metric. The main reason is labeling an account as PSM means it should be deleted. However, removing a real user is costly. Therefore, it is important to have a high precision to prevent removing real user.

3.4.1 Existing Method

Here we use the approach proposed by the top-ranked team in the DARPA Twitter Bot Challenge [12]. We consider all features that we could extract from our dataset. Our features include tweet syntax (average number of hashtags, average number of user mentions, average number of links, average number of special characters), tweet semantics (LDA topics), and user behavior (tweet spread, tweet frequency, tweet repeats). We apply three existing methods to detect PSM accounts: (1) *Random selection*: This method achieves the precision of 0.11. This also presents that our data is imbalanced and less than 12% of the users are PSM accounts. (2)

Table 3.1 Existing methods—false positive, true positive, precision, average cascade size, and median cascade size of the selected users as PSM

Method	FP	TP	Precision	Avg CS	Med CS
Random selection	80,700	10,346	0.11	**289.99**	**184**
Sentimetrix	640,552	77,984	0.11	261.37	171
Content based	292,039	43,483	0.13	267.66	174
Non-content based	357,027	63,025	0.15	262.97	172
All features	164,012	31,131	**0.16**	273.21	176

Sentimetrix: We cluster our data by DBSCAN algorithm. We then propagate the labels from 40 initial users to the users in each cluster based on the similarity metric. We use Support Vector Machines (SVM) to classify the remaining PSM accounts [12]. (3) *Classification* methods: In this experiment, we use the same labeled accounts as the previous experiment and apply different machine learning algorithms to predict the label of other samples. We group features based on the limitations of access to data into three categories. First, we consider only using content information (*Content based*) to detect the PSM accounts. Second, we use content independent features (*Non-content based*) [12] to classify users. Third, we apply all features (*All features*) to discriminate PSM accounts. The best result for each setting is when we apply Random Forest using all features. According to the results, this method achieves the highest precision of 0.16. Note that, most of the features used in the previous work and our baseline take advantage of both content and network structure. However, there are situations that the network information and content do not exist. In this situation, the best baseline has the precision of 0.15. We study the average (Avg CS) and median (Med CS) of the size of the cascades in which the selected PSM accounts have participated. Table 3.1 also illustrates the false positive, true positive, and precision of different methods.

3.4.2   Threshold-Based Selection Approach

In this experiment, we select all the users that satisfy the thresholds and check whether they are active or not. A user is *inactive*, if the account is suspended or closed. Since the dataset is not labeled, we label inactive users as PSM accounts. We set the threshold for all metrics to 0.7 except for relative likelihood causality (ϵ_{rel}), which is set to 7. We conduct two types of experiments: first, we study user selection for a given causality metric. We further study this approach using the combinations of metrics.

Single Metric Selection In this experiment, we attempt to select users based on each individual metric. As expected, these metrics can help us filter a significant amount of active users and overcome the data imbalance issue. Metric $\epsilon_{K\&M}$ achieves the largest recall in comparison with other metrics. However, it has the

Table 3.2 Threshold-based selection approach—false positive, true positive, precision, average cascade size, and median cascade size of the selected users using single metric

Method	FP	TP	Precision	Avg CS	Med CS
All features	164,012	31,131	**0.16**	273.21	176
Non-content based	357,027	63,025	0.15	262.97	172
$\epsilon_{K\&M}$	36,159	27,192	0.43	383.99	178
ϵ_{rel}	693	641	0.48	**567.78**	**211**
ϵ_{nb}	2,268	2,927	0.56	369.46	183.5
ϵ_{wnb}	7,463	14,409	**0.66**	311.84	164

Table 3.3 Threshold-based selection approach—number of common selected users using single metric

Status	Active			Inactive		
Method	ϵ_{rel}	ϵ_{nb}	ϵ_{wnb}	ϵ_{rel}	ϵ_{nb}	ϵ_{wnb}
$\epsilon_{K\&M}$	404	1,903	6,992	338	2,340	11,748
ϵ_{rel}		231	175		248	229
ϵ_{wnb}			1,358			1,911

largest number of false positives. Table 3.2 shows the performance of each metric. The precision value varies from 0.43 to 0.66 and metric ϵ_{wnb} achieves the best value. Metric ϵ_{rel} finds the more important PSM accounts with average cascade size of 567.78 and median of 211. In general, our detected PSM accounts have participated in the larger cascades in comparison with baseline methods.

We also observe that these metrics cover different regions of the search area. In other words, they select different user sets with little overlap between each other. The common users between any two pairs of the features are illustrated in Table 3.3. Considering the union of all metrics, 36,983 and 30,353 active and inactive users are selected, respectively.

Combination of Metrics Selection According to Table 3.3, most of the metric pairs have more inactive users in common than active users. In this experiment, we discuss if using the combination of these metrics can help improve the performance. We attempt to select users that satisfy the threshold for at least three metrics. We get 1,636 inactive users out of 2,887 selected ones, which works better than $\epsilon_{K\&M}$ and ϵ_{rel} while worse than ϵ_{nb} and ϵ_{wnb}. In brief, this approach achieves precision of 0.57. Moreover, the number of false positives (1,251) is lower than most of the other metrics.

3.4.3 Label Propagation Selection Approach

In label propagation selection, we first select a set of users that have a high causality score as seeds, then ProSel selects users that occur with those seeds and have a score higher than a threshold iteratively. Also, the seed set in each iteration is the selected users of the previous iteration. The intuition behind this approach is to select a user if it has a score higher than a threshold or has a lower score but occurs in the cascades that high score users occur. We set the parameters of *ProSel Algorithm* as follows: $\lambda = 0.1$, $\theta = 0.9$, except for relative likelihood causality, where we set $\lambda = 1$, $\theta = 9$. Table 3.4 shows the performance of each metric. Precision of these metrics varies from 0.47 to 0.75 and ϵ_{wnb} achieves the highest precision. Metrics ϵ_{rel} with average cascade size of 612.04 and ϵ_{nb} with median of 230 find the more important PSM accounts. Moreover, detected PSM accounts have participated in the larger cascades compared with threshold-based selection. This approach also produces much lower number of false positives compared to threshold-based selection. The comparison between this approach and threshold-based selection is illustrated in Fig. 3.2. From the precision perspective, label propagation method outperforms the threshold-based one.

Table 3.4 Label propagation selection approach—false positive, true positive, precision, average cascade size, and median cascade size of the selected users

Method	FP	TP	Precision	Avg CS	Med CS
All features	164,012	31,131	**0.16**	273.21	176
Non-Content based	357,027	63,025	0.15	262.97	172
$\epsilon_{K\&M}$	9,305	14,176	0.60	390.52	179
ϵ_{rel}	561	498	0.47	**612.04**	216
ϵ_{nb}	1,101	1,768	0.62	403.55	**230**
ϵ_{wnb}	1,318	4,000	**0.75**	355.24	183.5

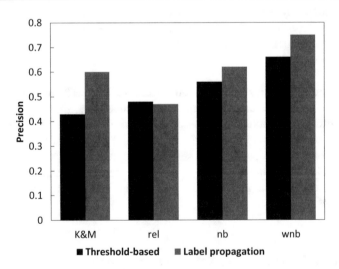

Fig. 3.2 Comparison between threshold-based and label propagation selection approaches for the inactive class

Table 3.5 Label propagation selection approach—number of common selected users

Status	Active			Inactive		
Method	ϵ_{rel}	ϵ_{nb}	ϵ_{wnb}	ϵ_{rel}	ϵ_{nb}	ϵ_{wnb}
$\epsilon_{K\&M}$	289	581	1,122	168	1,019	2,788
ϵ_{rel}		15	6		180	102
ϵ_{nb}			151			833

The number of common users selected by any pair of two metrics is also illustrated in Table 3.5. It shows that our metrics are powerful to cover different regions of the search area and identify different sets of users. In total, 10,254 distinct active users and 16,096 inactive ones are selected.

3.5 Conclusion

This chapter conducted a data-driven study on PSM accounts on ISIS-A dataset previously explained in Chap. 2. We proposed two unsupervised causality-based frameworks to detect these accounts. The proposed approaches outperformed the baselines. The advantages of these frameworks are: First, they are unsupervised. Second, they do not leverage any network structure, cascade path information, content, and user's information. We believe our technique can be applied in the areas such as detection of water armies and fake news campaigns.

References

1. H. Alvari, S. Hashemi, A. Hamzeh, Detecting overlapping communities in social networks by game theory and structural equivalence concept, in *International Conference on Artificial Intelligence and Computational Intelligence* (Springer, 2011), pp. 620–630
2. H. Alvari, A. Hajibagheri, G. Sukthankar, K. Lakkaraju, Identifying community structures in dynamic networks. Soc. Netw. Anal. Min. **6**(1), 77 (2016)
3. S. Baluja, R. Seth, D. Sivakumar, Y. Jing, J. Yagnik, S. Kumar, D. Ravichandran, M. Aly, Video suggestion and discovery for YouTube: taking random walks through the view graph, in *Proceedings of the 17th International Conference on World Wide Web* (ACM, 2008), pp. 895–904
4. G. Beigi, H. Liu, Similar but different: Exploiting users' congruity for recommendation systems, in *International Conference on Social Computing, Behavioral-Cultural Modeling, and Prediction* (Springer, 2018)
5. G. Beigi, J. Tang, H. Liu, Social science–guided feature engineering: A novel approach to signed link analysis. ACM Trans. Intell. Syst. Technol. **11**(1), 1–27 (Jan. 2020)
6. Q. Cao, M. Sirivianos, X. Yang, T. Pregueiro, Aiding the detection of fake accounts in large scale social online services, in *Proceedings of the 9th USENIX conference on Networked Systems Design and Implementation* (USENIX Association, 2012), pp. 15–15
7. C.A. Davis, O. Varol, E. Ferrara, A. Flammini, F. Menczer, BotOrNot: A system to evaluate social bots, in *Proceedings of the 25th International Conference Companion on World Wide Web* (International World Wide Web Conferences Steering Committee, 2016), pp. 273–274

8. J.P. Dickerson, V. Kagan, V. Subrahmanian, Using sentiment to detect bots on twitter: Are humans more opinionated than bots? in *2014 IEEE/ACM International Conference on Advances in Social Networks Analysis and Mining*, pp. 620–627 (2014)
9. F. Morstatter, L. Wu, T.H. Nazer, K.M. Carley, H. Liu, A new approach to bot detection: striking the balance between precision and recall, in *2016 IEEE/ACM International Conference on Advances in Social Networks Analysis and Mining*, pp. 533–540 (2016)
10. U.N. Raghavan, R. Albert, S. Kumara, Near linear time algorithm to detect community structures in large-scale networks. Phys. Rev. E **76**(3), 036106 (2007)
11. E. Shaabani, R. Guo, P. Shakarian, Detecting pathogenic social media accounts without content or network structure, in *2018 1st International Conference on Data Intelligence and Security (ICDIS)* (IEEE, 2018), pp. 57–64
12. V. Subrahmanian, A. Azaria, S. Durst, V. Kagan, A. Galstyan, K. Lerman, L. Zhu, E. Ferrara, A. Flammini, F. Menczer, The DARPA twitter bot challenge. Computer **49**(6), 38–46 (2016)
13. X. Zhang, J. Zhu, Q. Wang, H. Zhao, Identifying influential nodes in complex networks with community structure. Know. Based Syst. **42**, 74–84 (2013)
14. X. Zhu, Z. Ghahramani, Learning from labeled and unlabeled data with label propagation (2002)

Chapter 4
Early Detection of Pathogenic Social Media Accounts

4.1 Introduction

Early detection of PSMs in social media is crucial as they are likely to be *key* users to malicious campaigns [13]. This is a challenging task for several reasons. First, these platforms are primarily based on reports they receive from their own users to manually shut down PSMs which is not a timely approach. Despite efforts to suspend these accounts, many of them simply return to social media with different accounts. Second, the available data is often imbalanced and social network structure, which is at the core of many techniques [1, 2, 5, 6, 8, 14, 15], is not readily available. Third, PSMs often seek to utilize and cultivate large number of online communities of passive supporters to spread as much harmful information as they can.

In this chapter, causal inference [3, 4] from Chap. 2 is tailored to identify PSMs since they are key users in making a harmful message "viral"—where "viral" is defined as an order-of-magnitude increase. We propose *time-decay* causal metrics to distinguish PSMs from normal users within a *short* time around their activity. Our metrics alone can achieve high classification performance in identification of PSMs soon after they perform actions. Next, we pose the following research question: *Are causality scores of users within a community higher than those across different communities?* We propose a causal community detection-based classification method (C^2DC) that takes causality of users and the community structure of their action log.

We make the following major contributions:

- We enrich the causal inference framework of [9] and present *time-decay* extensions of the causal metrics in [11] for early identification of PSMs.
- We investigate the role of community structure in early detection of PSMs, by demonstrating that users within a community establish stronger causal relationships compared to the rest.

H. Alvari et al., *Identification of Pathogenic Social Media Accounts*, SpringerBriefs in Computer Science, https://doi.org/10.1007/978-3-030-61431-7_4

- We conduct a suit of experiments on ISIS-B dataset. Our metrics reached F1-score of 0.6 in identifying PSMs, half way their activity, and identified 71% of PSMs based on first 10 days of their activity, via supervised settings. The community detection approach achieved precision of 0.84 based on first 10 days of users activity; the misclassified accounts were identified based on their activity of 10 more days.

4.1.1   Decay-Based Causal Measures

As also mentioned in Chap. 2, causal inference framework was first introduced in [9]. Later, [11] adopted the framework and extended it to suite the problem of identifying PSMs. They extend the Kleinberg–Mishra causality ($\epsilon_{K\&M}$) to a series of causal metrics. To recap, we briefly explain them in the following discussion. Before going any further, $\epsilon_{K\&M}$ is computed as follows:

$$\epsilon_{K\&M}(i) = \frac{\sum_{j \in \mathbf{R}(i)}(p_{i,j} - p_{\neg i,j})}{|\mathbf{R}(i)|} \tag{4.1}$$

This metric measures how causal user i is, by taking the average of $p_{i,j} - p_{\neg i,j}$ over $\mathbf{R}(i)$. The intuition here is user i is more likely to be cause of message m to become viral than user j, if $p_{i,j} - p_{\neg i,j} > 0$. The work of [11] devised a suit of the variants, namely relative likelihood causality (ϵ_{rel}), neighborhood-based causality (ϵ_{nb}), and its weighted version (ϵ_{wnb}). Note that none of these metrics were originally introduced for *early* identification of PSMs. Therefore, we shall make slight modifications to their notations to adjust our temporal formulations, using calligraphic uppercase letters. We define $\mathscr{E}_{K\&M}$ over a given time interval I as follows:

$$\mathscr{E}_{K\&M}^{I}(i) = \frac{\sum_{j \in \mathscr{R}(i)}(\mathscr{P}_{i,j} - \mathscr{P}_{\neg i,j})}{|\mathscr{R}(i)|} \tag{4.2}$$

where $\mathscr{R}(i)$, $\mathscr{P}_{i,j}$, and $\mathscr{P}_{\neg i,j}$ are now defined over I. The authors in [11] mention that this metric cannot spot all PSMs. They define another metric, relative likelihood causality $\mathscr{E}_{rel}$, which works by assessing relative difference between $\mathscr{P}_{i,j}$, and $\mathscr{P}_{\neg i,j}$. We use its temporal version over I, $\mathscr{E}_{rel}^{I}(i) = \frac{\mathscr{S}(i,j)}{|\mathscr{R}(i)|}$, where $\mathscr{S}(i, j)$ is defined as follows and α is infinitesimal:

$$\mathscr{S}(i, j) = \begin{cases} \frac{\mathscr{P}_{i,j}}{\mathscr{P}_{\neg i,j} + \alpha} - 1, & \mathscr{P}_{i,j} > \mathscr{P}_{\neg i,j} \\ 1 - \frac{\mathscr{P}_{\neg i,j}}{\mathscr{P}_{i,j}}, & \mathscr{P}_{i,j} \leq \mathscr{P}_{\neg i,j} \end{cases} \tag{4.3}$$

Table 4.1 F1-score results for PSM accounts using each causal metric in [11]

Metric	F1-score				
	10%	20%	30%	40%	50%
$\mathscr{E}^I_{K\&M}$	0.41	0.42	0.45	0.46	0.49
$\mathscr{E}^I_{rel}$	0.3	0.31	0.33	0.35	0.37
$\mathscr{E}^I_{nb}$	0.49	0.51	0.52	0.54	0.55
$\mathscr{E}^I_{wnb}$	**0.51**	**0.52**	**0.55**	**0.56**	**0.59**

Two other neighborhood-based metrics were also defined in [11], whose temporal variants are computed over I as $\mathscr{E}^I_{nb}(j) = \frac{\sum_{i \in \mathscr{Q}(j)} \mathscr{E}^I_{K\&M}(i)}{|\mathscr{Q}(j)|}$, where $\mathscr{Q}(j) = \{i \mid j \in \mathscr{R}(i)\}$ is the set of all users that user j belongs to their related users sets. Similarly, the second metric is a weighted version of the above metric and is called weighted neighborhood-based causality and is calculated as $\mathscr{E}^I_{wnb}(j) = \frac{\sum_{i \in \mathscr{Q}(j)} w_i \times \mathscr{E}^I_{K\&M}(i)}{\sum_{i \in \mathscr{Q}(j)} w_i}$. This is to capture different impacts that users in $Q(j)$ have on user j. We apply a threshold-based selection approach that selects PSMs from normal users, based on a given threshold. Following [11], we use a threshold of 0.7 for all metrics except $\mathscr{E}^I_{rel}$ for which we used a threshold of 7 (Table 4.1).

4.2 The Proposed Framework

4.2.1 Leveraging Temporal Aspects of Causality

Previous causal metrics do not take into account time-decay effect. They assume a steady trend for computing causality scores. This is an unrealistic assumption, as causality of users may change over time. We introduce a generic decay-based metric. Our metric assigns different weights to different time points of a given time interval, inversely proportional to their distance from t (i.e., smaller distance is associated with higher weight). Specifically, it performs the following: it (1) breaks down the given time interval into shorter time periods using a sliding-time window, (2) deploys an exponential decay function of the form $f(x) = e^{-\alpha x}$ to account for the time-decay effect, and (3) takes average of the causality values computed over each sliding-time window. Formally, ξ^I_k is defined as follows, where $k \in \{K\&M, rel, nb, wnb\}$:

$$\xi^I_k(i) = \frac{1}{|\mathscr{T}|} \sum_{t' \in \mathscr{T}} e^{-\sigma(t-t')} \times \mathscr{E}^\Delta_k(i) \tag{4.4}$$

where σ is a scaling parameter of the exponential decay function, $\mathscr{T} = \{t' \mid t' = t_0 + j \times \delta, j \in \mathbb{N} \wedge t' \leq t - \delta\}$ is a sequence of sliding-time windows, and δ is a small fixed amount of time, which is used as the length of each sliding-time window

Fig. 4.1 An illustration of how decay-based causality works. To compute $\xi_k^I(i)$ over $I = [t_0, t]$, we use a sliding window $\Delta = [t' - \delta, t']$ and take the average between the resultant causality scores $e^{-\sigma(t-t')} \times \mathscr{E}_k^\Delta(i)$

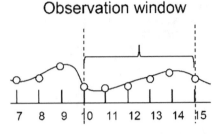

Observation window

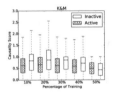
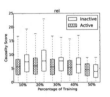
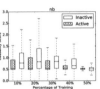
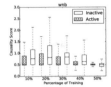

Fig. 4.2 Left to right: distributions of active and inactive users using ξ_k^I when $k \in \{K\&M, rel, nb, wnb\}$

Table 4.2 F1-score results for PSM accounts using each decay-based metric with and without communities

Metric	F1-score (without/with communities)				
	10%	20%	30%	40%	50%
$\xi_{K\&M}^I$	0.44/0.49	0.46/0.51	0.47/0.52	0.5/0.54	0.53/0.57
ξ_{rel}^I	0.36/0.4	0.38/0.43	0.39/0.46	0.41/0.49	0.42/0.5
ξ_{nb}^I	0.52/0.56	0.53/0.57	0.54/0.58	0.56/0.6	0.59/0.61
ξ_{wnb}^I	**0.54/0.57**	**0.55/0.58**	**0.57/0.6**	**0.58/0.62**	**0.6/0.63**

$\Delta = [t' - \delta, t']$ (Fig. 4.1). We depict in Fig. 4.2 box plots of the distributions of the decay-based metrics. To apply the threshold-based approach, we once again use a threshold of 0.7 for all metrics except ξ_{rel}^I for which we used a threshold of 7 (Table 4.2).

Early Detection of PSM Accounts *Given action log* **A**, *and user u where* $\exists t$ *s.t.* $(u, m, t) \in$ **A**, *our goal is to determine if u's account shall be suspended given its causality vector* $\mathbf{x}_u \in \mathbb{R}^d$ *(here, $d = 4$) computed using any of the causality metrics over* $[t - \delta, t + \delta]$.

4.2.2 Leveraging Community Structure Aspects of Causality

To answer the research question posed earlier, since network structure is not available, we need to build a graph $\mathbf{G} = (\mathbf{V}, \mathbf{E})$ from **A** by connecting any pairs of users who have posted same message *chronologically*. In this graph, **V** is a set of vertices (i.e., users) and **E** is a set of directed edges between users. For the sake of simplicity and without loss of generality, we make the edges of this graph undirected. Next, we leverage the LOUVAIN algorithm [7] to find the partitions $C = \{C_1, C_2, ..., C_k\}$ of k communities over $\mathbf{G}$. Among a myriad of the community detection algorithms [2, 10], we chose LOUVAIN due to its fast runtime and scalability—we leave examining other community detection algorithms to future work. Next, we perform the two-sample t-test $H_0 : v_a \geq v_b$, $H_1 : v_a < v_b$. The null hypothesis is: *users in a given community establish weak causal relations with each other as opposed to the other users in other communities*. We construct two vectors v_a and v_b as follows. We create v_a by computing Euclidean distances between causality vectors $(\mathbf{x}_i, \mathbf{x}_j)$ corresponding to each pair of users (u_i, u_j) who are from same community $C_l \in C$. Therefore, v_a contains exactly $\frac{1}{2} \sum_{l=1}^{|C|} |C_l|.(|C_l| - 1)$ elements. We construct v_b of size $\sum_{l=1}^{|C|} |C_l|$ by computing Euclidean distance between each user u_i in community $C_l \in C$, and a random user u_k chosen from the rest of the communities, i.e., $C \setminus C_l$. The null hypothesis is rejected at significance level $\alpha = 0.01$ with the p-value of 4.945e−17. We conclude that users in same communities are more likely to establish stronger causal relationships with each other than the rest of the communities. The answer to the question is thus positive. For brevity, we only reported results for 10% of the training set, while making similar arguments for other percentages is straightforward. Figure 4.3 shows box plots of the distributions of users using the decay-based metrics and the communities and same set of thresholds as before. We observe a clear distinction between active/suspended accounts, using the community structure. Results in Table 4.2 show improvements over previous ones.

First step of the proposed algorithm (Algorithm 2) involves finding the communities. In the second step, each unlabeled user is classified based on the available labels of her nearby peers in the same community. We use the K-NEAREST NEIGHBORS (KNN) algorithm to compute her k nearest neighbors in the same community, based on Euclidean distances between their causality vectors. We label her based on the majority class of her k nearest neighbors in the community. The merit of using community structure over merely using KNN is, communities can give finer-grained and more accurate sets of neighbors sharing similar causality scores.

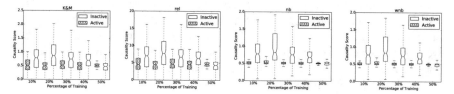

Fig. 4.3 Left to right: distributions of active and inactive users using communities and ξ_k^l when $k \in \{K\&M, rel, nb, wnb\}$

Algorithm 2 Causal community detection-based classification algorithm (C^2DC)

Require: Training samples $\{\mathbf{x}_1, ..., \mathbf{x}_N\}$, tests $\{\mathbf{x}'_1, ..., \mathbf{x}'_n\}$, $\mathbf{G}$, k
Ensure: Predicted labels $\{y'_1, ..., y'_n\}$
 1: $\mathbf{C} \leftarrow$ LOUVAIN($\mathbf{G}$)
 2: **for** each $\mathbf{x}'_i$ **do**
 3:      $C_l \leftarrow C' \in \mathbf{C}$ s.t. $\mathbf{x}'_i \in C'$
 4:      $\mathbf{D} \leftarrow \{\}$
 5:      **for** each $\mathbf{x}_j \in C_l$ **do**
 6:          $d_{ij} \leftarrow ||\mathbf{x}'_i - \mathbf{x}_j||_2$
 7:          $\mathbf{D} \leftarrow \mathbf{D} \cup \{d_{ij}\}$
 8:      **end for**
 9:      $\mathbf{K} \leftarrow$ KNN($\mathbf{D}, k$)
10:      $y'_i \leftarrow$ DOMINANT-LABEL($\mathbf{K}$)
11: **end for**

4.3   Experiments

We use different subsets of size $x\%$ of the entire time-line (from Feb 22, 2016 to May 27, 2016) of the action log $\mathbf{A}$, by varying x as $\{10, 20, 30, 40, 50\}$. For each subset and user i in the subset, we compute feature vector $\mathbf{x}_i \in \mathbb{R}^4$ of the corresponding causality scores. The feature vectors are then fed into supervised classifiers and the community detection-based algorithm. For the sake of fair comparison, we perform this for both causal and decay-based metrics. For both metrics, we empirically found that $\rho = 0.1$ and $\alpha = 0.001$ work well. For the decay-based causality metric we shall also assume a sliding window of size of 5 days (i.e., $\delta = 5$) and set $\sigma = 0.001$ which were found to work well in our experiments. Note we only present results for PSMs. Among many other supervised classifiers such as ADABOOST, LOGISTIC REGRESSION and SUPPORT VECTOR MACHINES (SVM), RANDOM FOREST (RF) with 200 estimators and "entropy" criterion, achieved the best performance. Therefore, for brevity we only report results when RF is used as the classifier.

We present results for the proposed community detection-based framework and causal and decay-based metrics. For computing k nearest neighbors, we set $k = 10$ as it was found to work well for our problem. By reporting the results of KNN trained on the decay-based causality features, we stress that using KNN alone does not yield a good performance. For the sake of fair comparison, all approaches were implemented and run in Python 2.7x, using the scikit-learn package. For any approach that requires special tuning of parameters, we conducted grid search to choose the best set of parameters.

4.3.1 Baseline Methods

4.3.1.1 CAUSAL [11]

We compare our metrics against the ones in [11] via supervised and community detection settings.

4.3.1.2 SENTIMETRIX-DBSCAN [12]

This was the winner of the DARPA challenge. It uses several features such as tweet syntax (e.g., average number of hashtags, average number of links), tweet semantics (e.g., LDA topics), and user behavior (e.g., tweet frequency). We perform 10-fold cross validation and use a held-out test set for evaluation. This baseline uses a seed set of 100 active and 100 inactive accounts, and then use DBSCAN clustering algorithm to find the associated clusters. Available labels are propagated to nearby unlabeled users in each cluster based on the Euclidean distance metric, and labels of the remaining accounts are predicted using SVM.

4.3.1.3 SENTIMETRIX-RF

This is a variant of [12] where we excluded the DBSCAN part and instead trained RF classifier using only the above features to evaluate the feature set.

4.3.2 Identification of PSM Accounts

For each subset a separate 10-fold cross validation was performed (Fig. 4.4). We observe the following:

- Community detection achieves the best performance using several metrics. This aligns well with the t-test results discussed earlier: *taking into account community structure of PSMs can boost the performance.*
- Causal and decay-based metrics mostly achieve higher performance than other approaches via both settings.
- Decay-based metrics are effective at identifying PSMs at different intervals via both settings. This lies at the inherent difference between decay-based and causal metrics—our metrics take into account time-decay effect.
- Although both variants of SENTIMETRIX-DBSCAN use many features, they were unable to defeat our approach.

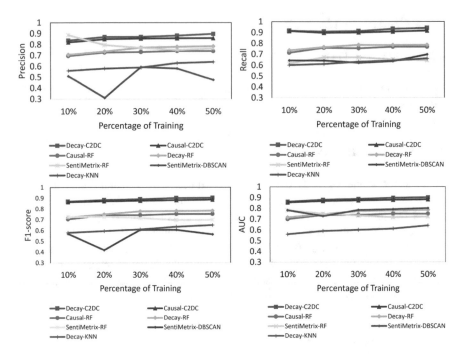

Fig. 4.4 Precision, recall, F1-score, and AUC results for each classifier. Experiments were run using 10-fold cross validation

4.3.3 Timeliness of PSM Accounts Identification

For each approach, we would like to see *how many* of PSMs who were active in the first 10 days of the dataset are correctly classified (i.e., true positives) as time goes by. Also, we need to keep track of false positives to ensure given approach does not merely label each instance as positive—otherwise a trivial approach that always label each instance as PSM would achieve highest performance. We are also interested to figure *how many* days need to pass to find these accounts. We train each classifier using 50% of the first portion of the dataset, and use a held-out set of the rest for evaluation. Next, we pass along the misclassified PSMs to the next portions to see how many of them are captured over time. We repeat the process until reaching 50% of the dataset—each time we increase the training set by adding new instances of the current portion.

There are 14,841 users in the first subset from which 3,358 users are PSMs. Table 4.3 shows the number of users from the first portion that (1) are correctly classified as PSM (out of 3,358), (2) are incorrectly classified as PSM (out of 29,617), over time. Community detection approaches were able to detect all PSMs who were active in the first 10 days of our dataset, no later than a month from their first activity. DECAY-C^{2}DC identified all of these PSMs in about 20 days since the

Table 4.3 True/false positives for PSM accounts. Numbers are out of 3,358/29,617 PSM/Normal accounts from the first period. Last column shows the number of PSM accounts from the first period which were not caught

Learner	True positives/false positives						Remaining
	02/22–03/02	03/02–03/12	03/12–03/22	03/22–03/31	03/31–04/09		
DECAY-C^2DC	3,072/131	286/0	0/0	0/0	0/0		0
CAUSAL-C^2DC	3,065/156	188/20	105/0	0/0	0/0		0
DECAY-KNN	2,198/459	427/234	315/78	109/19	96/0		213
DECAY-RF	2,472/307	643/263	143/121	72/68	28/0		0
CAUSAL-RF	2,398/441	619/315	221/169	89/70	51/0		0
SENTIMETRIX-RF	2,541/443	154/0	93/0	25/0	14/0		531
SENTIMETRIX-DBSCAN	2,157/2,075	551/5,332	271/209	92/118	72/696		215

first time they posted a message. Also, both causal and decay-based metrics when fed to RF classifier, identified all of the PSMs in the first period. SENTIMETRIX-DBSCAN and SENTIMETRIX-RF failed to detect all PSMs from the first portion, even after passing 50 days since their first activity. Furthermore, these two baselines generated much higher rates of false positives compared to the rest. The observations we make here are in line with the previous ones: *the proposed community detection-based framework is more effective and efficient than the rivals.*

4.4   Conclusion

We enriched the existing causal inference framework to suite the problem of early identification of PSMs. We proposed time-decay causal metrics which reached F1-score of 0.6 and via supervised learning identified 71% of the PSMs from the first 10 days of the dataset. We proposed a causal community detection-based classification algorithm, by leveraging community structure of PSMs and their causality. We achieved the precision of 0.84 for detecting PSMs within 10 days around their activity; the misclassified accounts were then detected 10 days later.

References

1. H. Alvari, S. Hashemi, A. Hamzeh, Detecting overlapping communities in social networks by game theory and structural equivalence concept, in *International Conference on Artificial Intelligence and Computational Intelligence* (Springer, 2011), pp. 620–630
2. H. Alvari, A. Hajibagheri, G. Sukthankar, K. Lakkaraju, Identifying community structures in dynamic networks. Soc. Netw. Anal. Min. **6**(1), 77 (2016)
3. H. Alvari, E. Shaabani, P. Shakarian, Early identification of pathogenic social media accounts. *IEEE Intelligent and Security Informatics* (2018). arXiv:1809.09331
4. H. Alvari, E. Shaabani, S. Sarkar, G. Beigi, P. Shakarian, Less is more: Semi-supervised causal inference for detecting pathogenic users in social media, in *Companion Proceedings of The 2019 World Wide Web Conference*, WWW '19 (Association for Computing Machinery, New York, NY, USA, 2019), pp. 154–161
5. G. Beigi, H. Liu, Similar but different: Exploiting users' congruity for recommendation systems, in *International Conference on Social Computing, Behavioral-Cultural Modeling, and Prediction* (Springer, 2018)
6. G. Beigi, J. Tang, H. Liu, Social science–guided feature engineering: A novel approach to signed link analysis. ACM Trans. Intell. Syst. Technol. **11**(1), 1–27 (Jan. 2020)
7. V.D. Blondel, J.-L. Guillaume, R. Lambiotte, E. Lefebvre, Fast unfolding of communities in large networks. J. Stat. Mech. Theory Exp. no. 10, P10008 (2008)
8. D. Kempe, J. Kleinberg, E. Tardos, Maximizing the spread of influence through a social network, in *KDD* (2003)
9. S. Kleinberg, B. Mishra, The temporal logic of causal structures. CoRR, abs/1205.2634 (2012)
10. A. Lancichinetti, F. Radicchi, J.J. Ramasco, S. Fortunato, Finding statistically significant communities in networks. PloS one **6**(4), e18961 (2011)

11. E. Shaabani, R. Guo, P. Shakarian, Detecting pathogenic social media accounts without content or network structure, in *2018 1st International Conference on Data Intelligence and Security (ICDIS)* (IEEE, 2018), pp. 57–64
12. V.S. Subrahmanian, A. Azaria, S. Durst, V. Kagan, A. Galstyan, K. Lerman, L. Zhu, E. Ferrara, A. Flammini, F. Menczer, The DARPA twitter bot challenge (2016)
13. O. Varol, E. Ferrara, F. Menczer, A. Flammini, Early detection of promoted campaigns on social media. EPJ Data Sci. **6**, 13 (2017)
14. L. Weng, F. Menczer, Y.-Y. Ahn, Predicting successful memes using network and community structure, in *ICWSM* (2014)
15. X. Zhang, J. Zhu, Q. Wang, H. Zhao, Identifying influential nodes in complex networks with community structure. Know. Based Syst. **42**, 74–84 (2013)

Chapter 5
Semi-Supervised Causal Inference for Identifying Pathogenic Social Media Accounts

5.1 Introduction

The problem of identification of PSMs has long been addressed in the past by the research community mostly in the form of bot detection. Several approaches especially supervised learning methods have been proposed in the literature and they have shown promising results [10]. However, for the most part, these approaches are often based on labeled data and exhaustive feature engineering. Examples of such feature groups include but are not limited to content, sentiment of posts, profile information, and network features. These approaches are thus very expensive as they require significant amount of efforts to design features and annotate large labeled datasets. In contrast, unlabeled data is ubiquitous and cheap to collect thanks to the massive user-generated data produced on a daily basis. Thus, in this work we set out to examine if unlabeled instances can be utilized to compensate for the lack of enough labeled data.

In this chapter, a semi-supervised causal inference [3] is tailored to detect PSMs who are promoters of misinformation online. We cast the problem of identifying PSMs as an optimization problem and propose a semi-supervised causal learning framework which utilizes unlabeled examples through manifold regularization [6]. In particular, we incorporate causality-based features extracted from users' activity log (i.e., cascades of retweets) as regularization terms into the optimization problem. In this work, causal inference is leveraged in an effort to capture whether or not PSMs exert causal influences while making a message viral. Similar to the previous chapters, the causality-based features here are built upon *Suppes' theory of probabilistic causation* [15] whose central concept is *prima facie causes*: an event to be recognized as a cause, must occur before the effect and must lead to an increase of the likelihood of observing the effect. While there exists a prolific literature on causality and their great impact in the computer-science community (see [12] for instance), we build our foundation on *Suppes' theory* as it is computationally less complex.

© The Author(s), under exclusive license to Springer Nature Switzerland AG 2021
H. Alvari et al., *Identification of Pathogenic Social Media Accounts*, SpringerBriefs in Computer Science, https://doi.org/10.1007/978-3-030-61431-7_5

The main contributions of this chapter are summarized below:

- We frame the problem of detecting PSM accounts as an optimization problem and present a Laplacian semi-supervised causal inference SEMIPSM for solving it. The unlabeled data are utilized via manifold regularization.
- Manifold regularization used in the resultant optimization formulation is built upon causality-based features created on a notion of *Suppes' theory of probabilistic causation*.
- We conduct a suite of experiments using different supervised and semi-supervised methods. Empirical experiments on a real-world ISIS-related dataset from Twitter suggest the effectiveness of the proposed semi-supervised causal inference over the existing methods.

5.2   The Proposed Method

In this section, we first provide the causal inference used to extract features out of users' activity log. Then, we detail the proposed semi-supervised causal inference, namely SEMIPSM, for detecting PSM accounts.

In this work, we use the time-decay causal metrics introduced in [2] and reviewed in Chap. 4 which are built on Suppes' theory. These metrics are fed to the final semi-supervised causal inference framework, as features. The first metric used in this work is $\mathscr{E}_{K\&M}$ which is computed over a given time interval I as follows:

$$\mathscr{E}^I_{K\&M}(i) = \frac{\sum_{j\in\mathscr{R}(i)}(\mathscr{P}_{i,j} - \mathscr{P}_{\neg i,j})}{|\mathscr{R}(i)|} \tag{5.1}$$

where $\mathscr{R}(i)$, $\mathscr{P}_{i,j}$, and $\mathscr{P}_{\neg i,j}$ are now defined over I. This metric estimates the causality score of user i in making a message *viral*, by taking the average of $\mathscr{P}_{i,j} - \mathscr{P}_{\neg i,j}$ over $\mathbf{R}(i)$. The intuition here is that user i is more likely to be a cause of message m to become viral than user j, if $\mathscr{P}_{i,j} - \mathscr{P}_{\neg i,j} > 0$. This metric cannot spot all PSMs, hence another metric is defined, namely relative likelihood causality $\mathscr{E}_{rel}$. This metric works by assessing relative difference between $\mathscr{P}_{i,j}$, and $\mathscr{P}_{\neg i,j}$:

$$\mathscr{E}^I_{rel}(i) = \frac{\mathscr{S}(i, j)}{|\mathscr{R}(i)|} \tag{5.2}$$

where $\mathscr{S}(i, j)$ is defined as follows and α is infinitesimal:

$$\mathscr{S}(i, j) = \begin{cases} \frac{\mathscr{P}_{i,j}}{\mathscr{P}_{\neg i,j}+\alpha} - 1, & \mathscr{P}_{i,j} > \mathscr{P}_{\neg i,j} \\ 1 - \frac{\mathscr{P}_{\neg i,j}}{\mathscr{P}_{i,j}}, & \mathscr{P}_{i,j} \leq \mathscr{P}_{\neg i,j} \end{cases} \tag{5.3}$$

Two other neighborhood-based metrics were also defined in [2], first of which is computed as

$$\mathcal{E}_{nb}^I(j) = \frac{\sum_{i \in \mathcal{Q}(j)} \mathcal{E}_{K\&M}^I(i)}{|\mathcal{Q}(j)|} \tag{5.4}$$

where $\mathcal{Q}(j) = \{i \mid j \in \mathcal{R}(i)\}$ is the set of all users that user j belongs to their related users sets. Similarly, the second metric is the weighted version of the above metric and is called weighted neighborhood-based causality and is calculated as

$$\mathcal{E}_{wnb}^I(j) = \frac{\sum_{i \in \mathcal{Q}(j)} w_i \times \mathcal{E}_{K\&M}^I(i)}{\sum_{i \in \mathcal{Q}(j)} w_i} \tag{5.5}$$

The aim of this metric is to capture different impacts that users in $Q(j)$ might have on user j.

5.2.1   Final Set of Features

Finally, the causal metrics discussed in the previous section will be fed as features to the semi-supervised framework—this will be described in the next section. The final set of features is in the following generic form ξ_k^I where $k \in \{K\&M, rel, nb, wnb\}$ [2] (see also Chap. 4):

$$\xi_k^I(i) = \frac{1}{|\mathcal{T}|} \sum_{t' \in \mathcal{T}} e^{-\sigma(t-t')} \times \mathcal{E}_k^\Delta(i) \tag{5.6}$$

Here, σ is a scaling parameter of the exponential decay function, $\mathcal{T} = \{t' \mid t' = t_0 + j \times \delta, j \in \mathbb{N} \wedge t' \le t - \delta\}$ is a sequence of sliding-time windows, and δ is a small fixed amount of time, which is used as the length of each sliding-time window $\Delta = [t' - \delta, t']$.

In essence, this metric assigns different weights to different time points of a given time interval, inversely proportional to their distance from t (i.e., smaller distance is associated with higher weight). Specifically, it performs the following: it (1) breaks down the given time interval into shorter time periods using a sliding-time window, (2) deploys an exponential decay function of the form $f(x) = e^{-\alpha x}$ to account for the time-decay effect, and (3) takes average of the causality values computed over each sliding-time window [2].

5.2.2   Semi-Supervised Causal Inference

Having defined the causality-based features, we now proceed to present the proposed semi-supervised Laplacian SVM framework, SEMIPSM. For the rest of the discussion, we shall assume a set of l labeled pairs $\{(x_i, y_i)\}_{i=1}^{l}$ and an unlabeled set of u instances $\{x_{l+i}\}_{i=1}^{u}$, where $x_i \in \mathbb{R}^n$ denotes the causality vector $\xi_k^l(i)$ of user i and $y_i \in \{+1, -1\}$ (PSM or not).

Recall for the standard soft-margin support vector machines, the following optimization problem is solved:

$$\min_{f_\theta \in \mathcal{H}_k} \gamma \|f_\theta\|_k^2 + C_l \sum_{i=1}^{l} H_1(y_i \, f_\theta(x_i)) \tag{5.7}$$

In the above equation, $f_\theta(\cdot)$ is a decision function of the form $f_\theta(\cdot) = w.\Phi(\cdot) + b$ where $\theta = (w, b)$ are the parameters of the model, and $\Phi(\cdot)$ is the feature map which is usually implemented using the kernel trick [7]. Also, the function $H_1(\cdot) = \max(0, 1 - \cdot)$ is the Hinge Loss function. The classical Representer theorem [6] suggests that solution to the optimization problem exists in a Hilbert space $\mathcal{H}_k$ and is of the form $f_\theta^*(x) = \sum_{i=1}^{l} \alpha_i^* \mathbf{K}(x, x_i)$. Here, $\mathbf{K}$ is the $l \times l$ Gram matrix over labeled samples. Equivalently, the above problem can be written as

$$\min_{w,b,\epsilon} \frac{1}{2} \|w\|_2^2 + C_l \sum_{i=1}^{l} \epsilon_i \tag{5.8}$$

$$s.t. \quad y_i(w.\Phi(x_i) + b) \geq 1 - \epsilon_i, \; i = 1, ..., l$$

$$\epsilon_i \geq 0, \; i = 1, ..., l \tag{5.9}$$

Next, we will use the above optimization equation as our basis to derive the formulations for our proposed semi-supervised learner.

The basic assumption behind semi-supervised learning methods is to leverage unlabeled instances in order to restructure hypotheses during the learning process [1]. Here, exogenous information extracted from causality-based features of users is exploited to make a better use of the unlabeled examples. To do so, we first introduce matrix $\mathbf{F}$ over both of the labeled and unlabeled samples with $\mathbf{F}_{ij} = \|\Phi(x_i) - \Phi(x_j)\|_2$ in $\|.\|_2$ norm. This way, we force instances x_i and x_j in our dataset to be relatively "close" to each other [4, 5], i.e., having a same label, if their corresponding causal-based feature vectors are close. To account for this, a regularization term is added to the standard equation and the following optimization is solved:

$$\min_{f_\theta \in \mathcal{H}_k} \frac{1}{2} \sum_{i=1}^{l} \mathbf{F}_{ij} \|f_\theta(x_i) - f_\theta(x_j)\|_2^2 = \mathbf{f}_\theta^T \mathcal{L}^T \mathbf{f}_\theta \tag{5.10}$$

where $\mathbf{f} = [f(x_1), ..., f(x_{l+u})]^T$ and $\mathcal{L}$ is the Laplacian matrix based on $\mathbf{F}$ given by $\mathcal{L} = \mathbf{D} - \mathbf{F}$, and $\mathbf{D}_{ii} = \sum_{j=1}^{l+u} \mathbf{F}_{ij}$. The intuition here is that causal pairs are more likely to have same labels than others.

Following the notations used in [6] and by including our regularization term, we would extend the standard equation by solving the following optimization:

$$\min_{f_\theta \in \mathcal{H}_k} \gamma \| f_\theta \|_k^2 + C_l \sum_{i=1}^{l} H_1(y_i f_\theta(x_i)) + C_r \mathbf{f}_\theta^T \mathcal{L} \mathbf{f}_\theta \tag{5.11}$$

Again, solution in $\mathcal{H}_k$ would be in the following form $f_\theta^*(x) = \sum_{i=1}^{l+u} \alpha_i^* \mathbf{K}(x, x_i)$. Here $\mathbf{K}$ is the $(l + u) \times (l + u)$ Gram matrix over all samples. The Eq. 5.11 could be then written as follows:

$$\min_{\alpha, b, \epsilon} \frac{1}{2} \alpha^T \mathbf{K} \alpha + C_l \sum_{i=1}^{l} \epsilon_i + \frac{C_r}{2} \alpha^T \mathbf{K} \mathcal{L} \mathbf{K} \alpha \tag{5.12}$$

$$s.t. \quad y_i \left(\sum_{j=1}^{l+u} \alpha_j \mathbf{K}(x_i, x_j) + b \right) \geq 1 - \epsilon_i, \quad i = 1, ..., l$$

$$\epsilon_i \geq 0, \quad i = 1, ..., l \tag{5.13}$$

With introduction of the Lagrangian multipliers β and γ, we write the Lagrangian function of the above equation as follows:

$$L(\alpha, \epsilon, b, \beta, \gamma) = \frac{1}{2} \alpha^T \mathbf{K}(I + C_r \mathcal{L}) \alpha + C_l \sum_{i=1}^{l} \epsilon_i \tag{5.14}$$

$$- \sum_{i=1}^{l} \beta_i \left(y_i \left(\sum_{j=1}^{l+u} \alpha_j \mathbf{K}(x_i, x_j) + b \right) - 1 + \epsilon_i \right) - \sum_{i=1}^{l} \gamma_i \epsilon_i$$

Obtaining the dual representation requires taking the following steps:

$$\frac{\partial L}{\partial b} = 0 \rightarrow \sum_{i=1}^{l} \beta_i y_i = 0 \tag{5.15}$$

$$\frac{\partial L}{\partial \epsilon_i} = 0 \rightarrow C_l - \beta_i - \gamma_i = 0 \rightarrow 0 \leq \beta_i \leq C_l \tag{5.16}$$

With the above equations, we formulate the reduced Lagrangian as a function of only α and β as follows:

$$L^R(\alpha, \beta) = \frac{1}{2}\alpha^T \mathbf{K}(I + C_r\mathcal{L})\alpha - \alpha^T \mathbf{K}\mathbf{J}^T \mathbf{Y}\beta + \sum_{i=1}^{l} \beta_i \qquad (5.17)$$

In the above equation, $\mathbf{J} = [\mathbf{I}\ \mathbf{0}]$ is a $l \times (l + u)$ matrix, $\mathbf{I}$ is the $l \times l$ identity matrix, and $\mathbf{Y}$ is a diagonal matrix consisting of the labels of the labeled examples. We first take the derivative of L^R with respect to α and then set $\frac{\partial L^R(\alpha,\beta)}{\partial \alpha} = 0$. We have the following equation:

$$\mathbf{K}(I + C_r\mathcal{L})\alpha - \mathbf{K}\mathbf{J}^T\mathbf{Y}\beta = 0 \qquad (5.18)$$

Accordingly, we obtain α^* by solving the following equation:

$$\alpha^* = (I + C_r\mathcal{L})^{-1}\mathbf{J}^T\mathbf{Y}\beta^* \qquad (5.19)$$

Next, we obtain the dual problem in the form of a quadratic programming problem by substituting α back in the reduced Lagrangian function Eq. 5.17:

$$\beta^* = \text{argmax}_{\beta \in \mathbb{R}^l} \ -\frac{1}{2}\beta^T \mathbf{Q}\beta + \sum_{i=1}^{l} \beta_i \qquad (5.20)$$

$$s.t. \quad \sum_{i=1}^{l} \beta_i y_i = 0$$

$$0 \le \beta_i \le C_l \qquad (5.21)$$

where $\beta = [\beta_1, ..., \beta_l]^T \in \mathbb{R}^l$ are the Lagrangian multipliers and $\mathbf{Q}$ is obtained as follows:

$$\mathbf{Q} = \mathbf{Y}\mathbf{J}\mathbf{K}(I + (C_r\mathcal{L})\mathbf{K})^{-1}\mathbf{J}^T\mathbf{Y} \qquad (5.22)$$

We summarize the proposed semi-supervised framework in Algorithm 1. Our optimization problem is very similar to the standard optimization problem solved for SVMs, hence we use a standard optimizer for SVMs to solve our problem.

Algorithm 3 Semi-supervised causal inference for PSM detection (SEMIPSM)

Require: $\{(x_i, y_i)\}_{i=1}^{l}, \{x_{l+i}\}_{i=1}^{u}, C_l, C_r$.
Ensure: Estimated function $f_\theta : \mathbb{R}^n \to \mathbb{R}$
1: Construct matrix $\mathbf{F}$ based on the causality-based features
2: Compute the corresponding Laplacian matrix $\mathcal{L}$.
3: Construct the Gram matrix over all examples using $\mathbf{K}_{ij} = k(x_i, x_j)$ where k is a kernel function.
4: Compute α^* and β^* using Eq. 5.19 and Eq. 5.20 and a standard QP solvers.
5: Compute function $f_\theta^*(x) = \sum_{i=1}^{l+u} \alpha_i^* \mathbf{K}(x, x_i)$

5.2.3 Computational Complexity

Here, we will explain the scalability of the algorithm in terms of big-$\mathcal{O}$ notation for both constituents of the proposed framework separately. For the first part of the approach, given a set of $\mathcal{A}$ cascades, and average number of $avg(\tau)$ users' actions (i.e., timestamps) in each cascade where $\tau \in \mathcal{A}$, the complexity of computing causality scores is $\mathcal{O}(|\mathcal{A}|.(avg(\tau))^2)$ (note on average there are $(avg(\tau))^2$ pairs of users in each cascade). For the second part, i.e., learning the semi-supervised algorithm, the most time-consuming part is calculating the inverse of a dense Gram matrix which leads to $\mathcal{O}((l + u)^3)$ complexity, where l and u are number of labeled and unlabeled instances [6].

5.3 Experiments

In this section we conduct experiments on the ISIS-B dataset and present results for several supervised and semi-supervised approaches. We first explain the dataset and provide some data analysis. Then, we will present the baseline methods. Finally, results and discussion are provided.

5.3.1 Baseline Methods

We compare the proposed method SEMIPSM against the following baseline methods. Note for all methods, we only report results when their best settings are used.

- **LABELSPREADING (RBF KERNEL) [17].** This is a graph inference-based label spreading approach with radial basis function (RBF) kernel.
- **LABEL SPREADING (KNN KERNEL) [17].** Similar to the previous approach with K-nearest neighbor (KNN) kernel.
- **LSTM [10].** The word-level LSTM approach here is similar to the deep neural network models used for sequential word predictions. We adapt the neural network to a sequence classification problem where the inputs are the vector of words in each tweet and the output is the predicted label of the tweet. We first use the word2vec [11] embedding pre-trained from a set of tweets similar to the data representation in our Twitter dataset.
- **ACCOUNT-LEVEL (RF CLASSIFIER) [10]** This approach uses the following features of the user profiles: *Statuses Count, Followers Count, Friends Count, Favorites Count, Listed Count, Default Profile, Geo Enables, Profile Uses Background Image, Verified, Protected.* We chose this method over Botometer [16] as it achieved comparable results with far less number of features ([16] uses over 1,500 features)(see also [8]). According to [10], we report the best results when Random Forest (RF) is used.

- **TWEET-LEVEL (RF CLASSIFIER) [10].** Similar to the previous baseline, this method uses only a handful of features extracted from tweets: *retweet count, reply count, favorite count, number of hashtags, number of URLs, number of mentions.* Likewise, we use RF as the classification algorithm.
- **SENTIMETRIX [14].** This approach was proposed by the top-ranked team in the DARPA Twitter Bot Challenge. We consider all features that we could extract from our dataset. Our features include tweet syntax (average number of hashtags, average number of user mentions, average number of links, average number of special characters), tweet semantics (LDA topics), and user behavior (tweet spread, tweet frequency, tweet repeats). The proposed approach starts with a small seed set and propagates the labels. Since we have enough labeled data for the training part, we use Random Forest as the learning approach.
- **C^2DC [2].** This approach uses time-decay causal community detection-based classification to detect PSM accounts [2]. For community detection, this approach uses Louvain algorithm.

5.3.2   Results and Discussion

All experiments were implemented in Python 2.7x and run on a machine equipped with an Intel(R) Xeon(R) CPU of 3.50 GHz with 200 GB of RAM running Linux. The proposed approach was implemented using CVXOPT[1] package. Furthermore, we split the whole dataset into 50% training and 50% test sets for all experiments. We report results in terms of F1-score in Tables 5.1 and 5.2. For any approach that requires special tuning of parameters, we conducted grid search to choose the best set of parameters. Specifically, for the proposed approach, we set the penalty parameter as $C_l = 0.6$ and the regularization parameter $C_r = 0.2$, and used linear kernel. For LABELSPREADING (RBF), the default value of $\gamma = 20$ was used and for LABELSPREADING (KNN), number of neighbors was set to 5. Also, for random forest we used 200 estimators and the "entropy" criterion was used. For computing k nearest neighbors in C^2DC, we set $k = 10$.

Furthermore for LSTM, we preprocessed the individual tweets in line with the steps mentioned in [13]. Since the content of the tweets are in Arabic, we replaced special characters that were present in the text with their Arabic counterparts if they were present. We used word vectors of dimensions 100 and deployed the skip-gram technique for obtaining the word vectors where the input is the target word, while the outputs are the words surrounding the target words. To model the tweet content in a manner that uses it to predict whether an account is PSM or not, we used Long Short-Term Memory (LSTM) models [9]. For the LSTM architecture, we used the first 20 words in the tokenized Arabic text of each tweet and use padding in situations where the number of tokens in a tweet is less than 20. We used 30 units

---

[1]http://cvxopt.org/.

Table 5.1 F1-score results of various methods on the labeled data. For semi-supervised learners, the size of the unlabeled data is fixed to 10% of the training set. The best performance is in bold

Learner	F1-score
SEMIPSM (CAUSAL FEATURES)	**0.94**
SEMIPSM (ACCOUNT-LEVEL FEATURES)	0.89
SEMIPSM (TWEET-LEVEL FEATURES)	0.88
LABELSPREADING (KNN/CAUSAL FEATURES)	0.89
LABELSPREADING (RBF/CAUSAL FEATURES)	0.88
ACCOUNT-LEVEL (RF CLASSIFIER)	0.88
TWEET-LEVEL (RF CLASSIFIER)	0.82
SENTIMETRIX	0.54
LSTM	0.41
C^2DC	0.4

in the LSTM architecture (many to one). The output of the LSTM layer was fed to a dense layer of 32 units with ReLU activations. We added dropout regularization following this layer to avoid overfitting and the output was then fed to a dense layer which outputs the category of the tweets.

We depict in Table 5.1 classification performance of all approaches on the labeled data. For the proposed framework SEMIPSM, we examine three sets of features (1) causality-based features, (2) account-level features [10], and (3) tweet-level features [10]. For the graph inference-based semi-supervised algorithms, i.e., LABELSPREADING (RBF) and LABELSPREADING (KNN), we only report results where causality-based features are used as they achieved best performance with them. As it is observed from the table, the best results in terms of F1-score belong to SEMIPSM where causality-based features are used. The runner-up is SEMIPSM with account-level features and the next best approach is SEMIPSM where tweet-level features are deployed. This clearly demonstrates the significance of using manifold regularization in the Laplacian semi-supervised framework over using other semi-supervised methods, LABELSPREADING (RBF) and LABELSPREADING (KNN).

We further note that the supervised classifier Random Forest using both of the account-level and tweet-level features and the whole labeled dataset achieve worse or comparable results to the semi-supervised learners. The fact that obtaining several tweet and account-level features is not trivial and do not necessarily lead to the best classification performance, motivates us to use semi-supervised algorithms which use less number of labeled examples, and yet achieve competing performance. We also obtain an F1-score of 0.41 when LSTM is used—the poor performance of this neural network model can be attributed to the raw Arabic text content. It suggests that the Arabic tokens as a representation might not be very informative about the category of accounts it has been generated from and some kind of weighting might be necessary before the LSTM module is used.

Also, Table 5.2 shows the classification performance of the semi-supervised approaches with causality-based features. The results are achieved using different portions of the unlabeled data, i.e., {10%, 20%, 30%, 40%, 50%} of the training set. As it is seen in the table, SEMIPSM achieves the best performance on different portions of the unlabeled data compared to the other semi-supervised learners,

Table 5.2 F1-score results of the semi-supervised approaches when causality-based features are used. Results are reported on different portions of the unlabeled data. The best performance is in bold

	Percentage of unlabeled data				
	10%	20%	30%	40%	50%
SEMIPSM	**0.94**	**0.93**	**0.91**	**0.9**	**0.88**
LABELSPREADING (KNN)	0.89	0.88	0.87	0.85	0.81
LABELSPREADING (RBF)	0.88	0.86	0.85	0.82	0.80

while performances of all approaches deteriorate with increasing the percentage of the unlabeled data. Furthermore, SEMIPSM still outperforms all other supervised methods as well as LSTM and C^2DC when up to 50% of the data has been made unlabeled.

Observations Overall, we make the following observations:

- Among the semi-supervised learners used in this study, SEMIPSM achieves the best classification performance suggesting the significance of using unlabeled instances in the form of manifold regularization. Manifold regularization is shown effective in boosting the classification performance, with three different sets of features confirming this.
- Causality-based features achieve the best performance via both Laplacian and graph inference-based semi-supervised settings. This lies at the inherent property of the causality-based features—they are designed to show whether or not user i exerts a causal influence on j. This is effective in capturing PSMs as they are key users in making a message viral.
- Compared to the supervised methods ACCOUNT-LEVEL (RF) and TWEET-LEVEL (RF), semi-supervised learners achieve either comparable or best results, suggesting promising results with less number of labeled examples.
- Among the supervised methods ACCOUNT-LEVEL (RF) and TWEET-LEVEL (RF), the former achieves higher F1-score indicating that account-level features are more useful in boosting the performance, although they are harder to obtain [10].
- Semi-supervised learners achieve best or comparable results with supervised learners, even with up to 50% of the data made unlabeled. This clearly shows the superiority of using unlabeled examples over labeled ones.

5.4 Conclusion

In this chapter, we presented a semi-supervised Laplacian SVM to detect PSM users in social media who are promoters of misinformation spread. We cast the problem of identifying PSMs as an optimization problem and introduced a Laplacian semi-supervised SVM via utilizing unlabeled examples through manifold regularization.

We examined different sets of features extracted from users activity log (in the form of cascades of retweets) as regularization terms: (1) causality-based features and (2) LSTM-based features. Our causality-based features were built upon *Suppes' theory of probabilistic causation*. The LSTM-based features were extracted via LSTM which has shown promising results for different tasks in the literature.

References

1. H. Alvari, P. Shakarian, J.K. Snyder, Semi-supervised learning for detecting human trafficking. Security Informatics **6**(1), 1 (2017)
2. H. Alvari, E. Shaabani, P. Shakarian, Early identification of pathogenic social media accounts. *IEEE Intelligent and Security Informatics* (2018). arXiv:1809.09331
3. H. Alvari, E. Shaabani, S. Sarkar, G. Beigi, P. Shakarian, Less is more: Semi-supervised causal inference for detecting pathogenic users in social media, in *Companion Proceedings of The 2019 World Wide Web Conference*, pp. 154–161 (2019)
4. G. Beigi, H. Liu, Similar but different: Exploiting users' congruity for recommendation systems, in *International Conference on Social Computing, Behavioral-Cultural Modeling, and Prediction* (Springer, 2018)
5. G. Beigi, J. Tang, H. Liu, Social science-guided feature engineering: A novel approach to signed link analysis. ACM Trans. Intell. Syst. Technol. **11**(1), 1–27 (Jan. 2020)
6. M. Belkin, P. Niyogi, V. Sindhwani, Manifold regularization: A geometric framework for learning from labeled and unlabeled examples. J. Mach. Learn. Res. **7**(Nov), 2399–2434 (2006)
7. C. Cortes, V. Vapnik, Support-vector networks. Machine Learning **20**(3), 273–297 (1995)
8. E. Ferrara, O. Varol, C. Davis, F. Menczer, A. Flammini, The rise of social bots. Commun. ACM **59**(7), 96–104 (2016)
9. S. Hochreiter, J. Schmidhuber, Long short-term memory. Neural Computation **9**(8), 1735–1780 (1997)
10. S. Kudugunta, E. Ferrara, Deep neural networks for bot detection. Preprint (2018). arXiv:1802.04289
11. T. Mikolov, I. Sutskever, K. Chen, G.S. Corrado, J. Dean, Distributed representations of words and phrases and their compositionality, in *Advances in Neural Information Processing Systems*, pp. 3111–3119 (2013)
12. J. Pearl, *Causality: Models, Reasoning and Inference*, 2nd edn. (Cambridge University Press, New York, NY, USA, 2009)
13. A.B. Soliman, K. Eissa, S.R. El-Beltagy, Aravec: A set of Arabic word embedding models for use in Arabic NLP. Procedia Comput. Sci. **117**, 256–265 (2017)
14. V.S. Subrahmanian, A. Azaria, S. Durst, V. Kagan, A. Galstyan, K. Lerman, L. Zhu, E. Ferrara, A. Flammini, F. Menczer, The DARPA twitter bot challenge (2016)
15. P. Suppes, A probabilistic theory of causality (1970)
16. O. Varol, E. Ferrara, C.A. Davis, F. Menczer, A. Flammini, Online human-bot interactions: Detection, estimation, and characterization, in *ICWSM* (2017)
17. D. Zhou, O. Bousquet, T.N. Lal, J. Weston, B. Schölkopf, Learning with local and global consistency, in *Advances in Neural Information Processing Systems*, pp. 321–328 (2004)

Chapter 6
Graph-Based Semi-Supervised and Supervised Approaches for Detecting Pathogenic Social Media Accounts

6.1 Introduction

As we discussed in the previous chapters, the network structure is not always available [1, 2, 4, 5]. For example, the Facebook API does not make this information (i.e., friendship network) available without the permission of the users. Moreover, the use of content often necessitates the training of a new model for the previously unobserved topics. For example, PSM accounts taking part in elections in the USA and Europe will likely leverage different types of content. To deal with these issues, causal inference is tailored to detect PSM accounts in [7] based on an unsupervised learning method. In this chapter, expanding upon the work of [7] (see also Chap. 3), we propose causal-based framework that incorporates graph-based metrics to distinguish PSMs from normal users within a short time around their activities [8]. Our new metrics combined with our causal ones from Chap. 3, can achieve high precision of 0.90, while increasing the recall from 0.22 to 0.49. We propose supervised and semi-supervised approaches and then show our proposed methods outperform the ones in the literature. In summary, the major contributions of this study are itemized as follows:

- We propose supervised and semi-supervised PSM detection frameworks that do not leverage network structure, cascade path information, content, and user's information.
- We introduce graph-based framework using the cascades and propose a series of scalable metrics to identify PSM users. We apply this framework to more than 722K users and 35K cascades.
- We propose a deep neural network framework which achieves AUC of 0.82. We show that our framework significantly outperforms Sentimetrix [9] (0.74), causality [7] (0.73), time-decay causality [3] (0.66), and causal community detection-based classification [3] (0.6).

© The Author(s), under exclusive license to Springer Nature Switzerland AG 2021
H. Alvari et al., *Identification of Pathogenic Social Media Accounts*, SpringerBriefs in Computer Science, https://doi.org/10.1007/978-3-030-61431-7_6

- We introduce a self-training semi-supervised framework that can capture more than 29K PSM users with the precision of 0.81. We only used 600 labeled data for training and development sets. Moreover, if a supervisor is involved in the training loop, the proposed algorithm is able to capture more than 80K PSM users.

6.2   Technical Approach

6.2.1   Graph-Based Framework

User–Message Bipartite Graph   Here, we denote $Actions$ as a bipartite graph $G_{u-m}(U, M, E)$, where users U and messages M are disjoint sets of vertices. There is an annotated link from user u to message m if u has tweeted/retweeted m and is annotated by occurrence time t (see Fig. 6.1). In other words, every edge in graph G_{u-m} is associated with one tuple $(u, m, t) \in Actions$. For a given node $u \in U (m \in M)$, the set $\mathcal{N}_u = \{m' \in M \ s.t. \ (u, m') \in E\}(\mathcal{N}_m = \{u' \in U \ s.t. \ (u', m) \in E\})$ is the set of immediate neighbors of u (m). We also define $U^v \subset U$ which is the set of verified users (often celebrities). We indicate $U_m^v = \{u | (u, m, t) \in Actions, u \in U^v\}$ as a set of verified users that have re/tweeted message m.

As for the edges, we examine different metrics such as Jaccard similarity between users, and rank of a user in a message which is defined as $Rank_{(u,m)} = |\{(u', m, t') \in Actions | (u, m, t) \in Actions, t' < t\}|$. We also define normalized rank as

$$NR(u, m) = 1 - \frac{Rank_{(u,m)}}{\mathcal{N}_m} \tag{6.1}$$

Our intuition behind rank metric $Rank_{(u,m)}$ is that the earlier a user has participated in spreading a message, the more important the user is. In this regard, we can also define the exponential decay of the time as

$$\mathcal{T}_u^m = \exp(-\gamma \Delta t_u^m), \tag{6.2}$$

where $\Delta t_u^m = \{t | (u, m, t) \in Actions\} - \min(\{t' | (u', m, t') \in Actions\})$, and γ is a constant. This metric prioritizes based on the retweeting time of the message. In other words, this metric assigns different weights to different time points of a given time interval, inversely proportional to their duration from start of the cascade, i.e., smaller duration is associated with higher weight.

Using all these information, we then annotated users U based on their local and network characteristics such as degree, and PageRank. We also consider function $\mathcal{F} \in \{sum, max, min, avg, med, std\}$ to calculate statistics such as minimum,

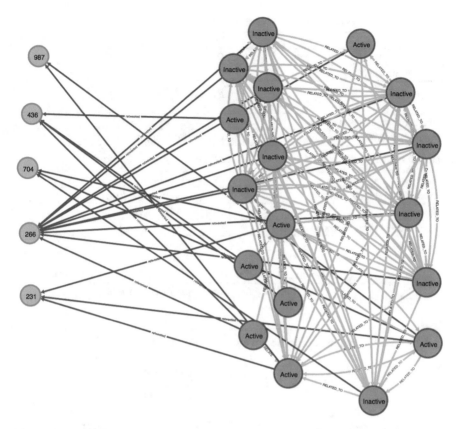

Fig. 6.1 User-message bipartite graph and user graph. Red and purple nodes represent users and messages, respectively. Users are labeled by their current status (Active, Not Found, Suspended) and messages with the length of the cascade (degree). Blue and green edges represent user–message and user–user relationships

mean, median, maximum, and standard deviation based on their one-hop or two-hops neighbors. For example, for a given user u, mean of re/tweeted message's PageRank of user u proved to be among top predictive metrics according to our experiments. Using these intuitions, we explored the space of variants features and list those we found to be best-performing in Table 6.1.

User Graph We represent a directed weighted user graph $G(V', E')$ where the set of nodes V' corresponds with key users. There is a link between two users if they are both key users of at least a message. There is a link from i (j) to j (i) if the number of times that "i appears before j and both are key users" is equal to or larger (smaller) than the case when "j appears before i," see Fig. 6.1. For a given node i, the set $N_i^{out} = \{i' \in V' \; s.t. \; (i, i') \in E'\}$ $(N_i^{in} = \{i' \in V' \; s.t. \; (i', i) \in E'\})$- the set of outgoing (incoming) immediate neighbors of i. The weight of edges is determined as a variant of co-occurrences of the key user pairs:

Table 6.1 User–message bipartite graph-based metrics

Name	Definition
Degree	$D_v = |\{v'|(v, v') \in E \vee (v', v) \in E\}|$
Cascade size statistics	$CS_{u,\mathscr{F}} = \mathscr{F}_{m \in \mathscr{N}_u} D_m$
PageRank	$PR(v) = \frac{1-d}{N} + d \sum_{v' \in \mathscr{N}_v} \frac{PR(v')}{L(v')}$
Message's PageRank statistics	$PS_{u,\mathscr{F}} = \mathscr{F}_{m \in \mathscr{N}_u} PR(m)$
Number of verified users	$Vr_m = |\{u|(u, m) \in E, u \in U^v\}|$
Jaccard similarity statistics	$JS_{u,\mathscr{F}} = \mathscr{F}_{u' \in U} \frac{|\mathscr{N}_u \cap \mathscr{N}_{u'}|}{|\mathscr{N}_u \cup \mathscr{N}_{u'}|}$
Intersection statistics	$IS_{u,\mathscr{F}} = \mathscr{F}_{u' \in U} |\mathscr{N}_u \cap \mathscr{N}_{u'}|$
Normalized rank statistics	$NRS_{u,\mathscr{F}} = \mathscr{F}_{m \in \mathscr{N}_u} NR(u, m)$
$\mathscr{T}$ statistics	$\mathscr{T}S_{u,\mathscr{F}} = \mathscr{F}_{m \in \mathscr{N}_u} \mathscr{T}_u^m$
Verified users in the cascades statistics	$U^v S_{u,\mathscr{F}} = \mathscr{F}_{m \in \mathscr{N}_u} |U_m^v|$

$$\mathscr{CO}_{i,j} = \frac{|\{m|i, j \text{ are key users, } \exists t, t' \text{ where } t < t', (i, m, t), (j, m, t') \in Actions\}|}{\min(|\{m|i \text{ is a key user}\}|, |\{m|j \text{ is a key user}\}|)} \tag{6.3}$$

Using $\mathscr{CO}_{i,j}$, we then propose a weighted co-occurrence score for user i as

$$\mathscr{CO}_{i,N_i}^w = \frac{\sum_{j \in N_i} (abs(\delta_{i,j}) + 1) \times \mathscr{CO}_{i,j}}{\sum_{j \in N_i} (abs(\delta_{i,j}) + 1)} \tag{6.4}$$

where $abs(\cdot)$ denotes the absolute value of the input. The differences between ordered joint occurrences $\delta_{i,j}$ is also defined as

$$\delta_{i,j} = |\{m|\exists t, t' \text{ s.t. } t < t', (i, m, t), (j, m, t') \in Actions\}| \tag{6.5}$$
$$-|\{m|\exists t, t' \text{ s.t. } t > t', (i, m, t), (j, m, t') \in Actions\}|$$

The list of user graph-based metrics extracted from graph G is shown in Table 6.2. We further calculate the probability of "user j appears after user i" as

$$P_{(j,i)} = \frac{|\{m \in M_{vir}|\exists t, t' \text{ where } t < t' \text{ and } (i, m, t), (j, m, t') \in Actions\}|}{|\{m|(j, m, t) \in Actions\}|} \tag{6.6}$$

The average probability that user i appears before its related users $R(i)$ is also a good indicator for identifying PSM accounts:

$$CM_i = \frac{\sum_{R(i)} P_{(j,i)}}{|R(i)|} \tag{6.7}$$

We aim to evaluate users from different perspectives and these metrics have shown to be helpful for evaluating users and detecting PSM accounts.

Table 6.2 User graph-based metrics

Description	Definition
Degree	$|N_i^{out}|$
Outgoing co-occurrence score statistics	$\mathcal{C}\mathcal{O}S_{i,\mathcal{F}}^{out} = \mathcal{F}_{j \in \mathcal{N}_i^{out}} \mathcal{C}\mathcal{O}_{i,j}$
Incoming co-occurrence score statistics	$\mathcal{C}\mathcal{O}S_{i,\mathcal{F}}^{in} = \mathcal{F}_{j \in \mathcal{N}_i^{in}} \mathcal{C}\mathcal{O}_{i,j}$
Weighted co-occurrence score	$\mathcal{C}\mathcal{O}_{i,N_i^{out}}^{w}$
Number of outgoing verified users	$|\{j | j \in N_i^{out}, j \in U^v\}|$
Number of incoming verified	$|\{j | j \in N_i^{in}, j \in U^v\}|$
Triangles	Number of triangles
Clustering coefficient	$CC_i = \frac{|\{(j,k)|j,k \in N_i,(j,k) \in E'\}|}{|N_i| \times (|N_i|-1)}$

6.2.2   Problem Statement

Our goal is to find the potential PSM accounts from the cascades. In the previous section, we discussed causality metrics, and defined diverse set of features using both user–message bipartite and user graphs where these metrics can discriminate the users of interest.

Problem (Early PSM Account Detection) *Given Action log Actions, causality and structural metrics, we wish to identify set of key users that are PSM accounts.*

6.3   PSM Account Detection Algorithm

We employ supervised, and semi-supervised approaches for detecting PSM accounts. Proposed metrics are scalable and can be calculated efficiently using map-reduce programming model and storing data in a graph-based database. To such aim, we used Neo4j to store data and calculated most of the structural metrics using Cypher query language [10].

6.3.1   Supervised Learning Approach

We evaluate several supervised learning approaches including logistic regression (LR), Naive Bayes (NB), k-nearest neighbors (KNN), and random forest (RF) on the same set of features. We also develop a dense deep neural network structure using Keras. As for the deep neural network and in order to find the best architecture and hyperparameters, we utilize the random search method. Many model structures were tested and Fig. 6.2 illustrates the best architecture.

As we can see from Fig. 6.2, the proposed deep neural net, in fact, consists of three dense deep neural net structures. The first two structures are of the same, but

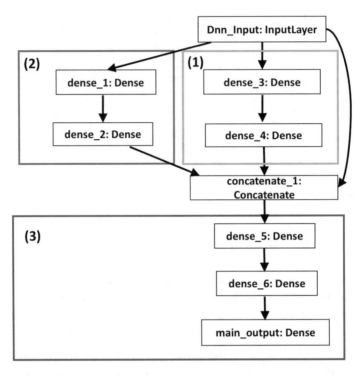

Fig. 6.2 The proposed deep neural net structure

the activation functions of their layers are different. The intuition is that we aimed to capture the most useful information from the input data and our experiments show the ReLU and Sigmoid activation functions can contribute to this. Specifically, these two structures are aimed to filter the noises in the input data and prepare clean inputs to feed into the third structure. In this regard, the outputs of these two structures along with the input data are concatenated into one vector and this vector is fed into another dense deep neural net. Finally, the output of this structure is fed into a regular output layer. To avoid overfitting, we used dropout method. In the proposed framework, the binary cross entropy loss function is minimized and the best optimizers are reported as Adam and Adagrad.

6.3.2 Self-Training Semi-Supervised Learning Approach

Semi-supervised algorithms [9, 11] use unlabeled data along with the labeled data to better capture the shape of the underlying data distribution and generalize better to new samples. Here, we propose a *Weighted Self-Training Algorithm* (WSET) shown in Algorithm 4 to address such problem. We start with small amount of labeled

---

Algorithm 4 Weighted self-training algorithm (WSET)

1: **procedure** WSET($L = \{\mathbf{u}_i, l_i\}, U = \{\mathbf{u}_j\}, \alpha, \beta, \theta_{pr}, \theta_{tr}$)
2:     Split L to training set L_t and development set L_d
3:     $L_t.w_c = 1$
4:     $it = 1$
5:     $m = $ Train a classification model using L_t
6:     $L_d.p = $ confidence score p using m of L_d
7:     $c = $ accuracy of model m on L_d
8:     $c' = c$
9:     **while** $c' >= c - \theta_{tr}$ **do**
10:         $U.p = $ confidence score p using m of U
11:         Update L_t and U by Algorithm 5 ($L_t, L_d, U, \alpha, \beta, \theta_{pr}, it$)
12:         $m = $ Train a classification model using L_t
13:         $L_d.p = $ confidence score p using m of L_d
14:         $c' = $ accuracy of model m on L_d
15:         $it = it + 1$
16:     **end while**
17:     **return** L_t
18: **end procedure**

---

training data and iteratively add users with high confidence scores from unlabeled data to the training set. Let us denote labeled data $L = \{\mathbf{u}_i, l_i\}$ and unlabeled data $U = \{\mathbf{u}_j\}$. Labeled data is split into training set L_t and development set L_d. We then iteratively train a classifier using training set and predict the confidence scores for development set and unlabeled data. Based on the *confidence score* obtained from *development set*, a *threshold* is determined. We then select *all samples* from *unlabeled data* that *satisfy the threshold*. Next, those samples are *removed* from *unlabeled set* and are *added* to the *training set*. The termination condition is determined based on at most θ_{tr} drop in accuracy on the development set or minimum number of selected users by algorithm.

There are still two main questions that need to be answered:

RQ1. Should all training samples be weighted equally?
RQ2. How should a threshold be determined for adding unlabeled data to the labeled set?

Since the prediction mistake reinforces itself, and the prediction error increases by number of iterations, the way we choose samples is of importance. According to our experiments, all training samples should not be weighted equally. We found the *exponential decay weighting* approach as the most efficient one (see **RQ1**). Considering a sample with confidence p_l associated to a specific label l in iteration it, the exponential decay weighting approach is defined as

$$\exp\left(-\beta \times it \times \left(\frac{1}{1 - p_l}\right)\right) \tag{6.8}$$

where β is a parameter. To answer the second question, we pick the threshold to have the minimum precision of θ_{pr} on development set in each iteration. Since

---

Algorithm 5 Update weighted self-training datasets algorithm (UPDWSET)

1: **procedure** UPDWSET($L_t, L_d, U, \alpha, \beta, \theta_{pr}, it$)
2:     $S = \emptyset$
3:     **for** $l \in [True, False]$ **do**
4:         $thr = Find Precision Threshold(L_d, \theta_{pr} - \alpha \times (it - 1), label = l)$
5:         $S = S \cup \{\mathbf{u} \in U | \mathbf{u}.p \geq thr\}$
6:     **end for**
7:     $U = U - S$
8:     $S.w_c = \exp(-\beta \times it \times (\frac{1}{1-p}))$
9:     $L_t = L_t \cup S$
10:     **return** L_t, U
11: **end procedure**

---

the precision decreases as the algorithm iterates, the threshold is required to be adjusted in order to make sure the top-ranked and qualified samples are picked up. Mathematically, the updated threshold in each iteration is defined as follows:

$$\theta_{pr} - \alpha \times (it - 1) \tag{6.9}$$

where α is a parameter, $\alpha \in [0, \frac{1}{it-1}], it > 0$. We pick 0.005 for the experiments. If $it = 1$, the threshold is equal to θ_{pr}. As the number of iteration increases the threshold is updated according to the product of α and iteration number it. This approach can make sure that we are picking samples with acceptable confidence. Algorithm 5 presents our approach for updating labeled and unlabeled datasets.

6.4   Results and Discussion

We implement part of our code in Scala Spark and Python 2.7x and run it on a machine equipped with an Intel Xeon CPU (2 processors of 2.4 GHz) with 256 GB of RAM running Windows 7. We also implement most of structural metrics in Cypher query language. We create the graphs using Neo4j [10] on a machine equipped with an Intel Xeon CPU (2 processors of 2.4 GHz) with 520 GB of RAM. We set the parameter ϕ as 0.5 to label key users (Definition 2.2). That is, we are looking for the users that participate in the *action* before the number of participants gets twice.

In the following sections, first we look at the baseline methods. Then we address the performance of two proposed approaches (see Sect. 6.3): (1) *Supervised Learning Approach*: applying different supervised learning methods on proposed metrics, (2) *Self-Training Semi-Supervised Learning Approach*: selecting users by applying Algorithm 4. The intuition behind this approach is to select users with the high probability of being either PSM or non-PSM (normal user) from unlabeled data and then adding them to the training set in order to improve the performance. We evaluate methods based on both Precision-Recall and Receiver

Table 6.3 Statistics of the datasets used in experiments

Name	PSM accounts	Normal accounts	Total
$\mathscr{A}$	19,859	65,417	85,276
$\mathscr{B}$	137,248	585,396	722,644

Operating Characteristics (ROC) curves. Note that in all experiments, the training, development, and test sets are imbalanced with more normal users than PSM users. The statistics of the datasets built from the ISIS-A data (see Chap. 2) are presented in Table 6.3. Dataset $\mathscr{A}$ is randomly selected from dataset $\mathscr{B}$ using sklearn library [6]. Note that, all random selections of data in the experiments have been done using sklearn library. We repeated the experiments 3 times and picked the median output. It is worth to mention that the variance among the results was negligible. In this problem, our goal is to achieve high precision while maximizing the recall. The main reason is labeling an account as PSM means it should be deleted. However, removing a normal user is costly. Therefore, it is important to have a high precision to prevent removing the normal users.

6.4.1 Baseline Methods

We have compared our results with existing work for detecting PSM accounts [3, 7] or bots [9].

Causality This work presents a set of causality metrics and unsupervised label propagation model to identify PSM accounts [7]. However, since our approach is supervised, we only use the causality metrics and evaluate its performance in a supervised framework.

C^{2}DC This approach uses time-decay causal community detection-based classification to detect PSM accounts [3]. We also considered time-decay causal metrics with random forest as another baseline method (TDCausality).

Sentimetrix This approach is proposed by the top-ranked team in the DARPA Twitter Bot Challenge [9]. We consider all features that we could extract from our dataset. Our features include tweet syntax (average number of hashtags, average number of user mentions, average number of links, average number of special characters), tweet semantics (LDA topics), and user behavior (tweet spread, tweet frequency, tweet repeats). The proposed method starts with a small seed set and propagate the labels. As we have enough labeled dataset for the training set, we use random forest as the learning approach.

We use dataset $\mathscr{A}$ to evaluate different approaches. Figure 6.3 (left) shows the precision-recall curve for these methods. As it is shown, in the supervised framework, *Sentimetrix* outperforms all approaches in general. Also, *Causality* is a comparable approach with Sentimetrix with the constraint that the precision is no

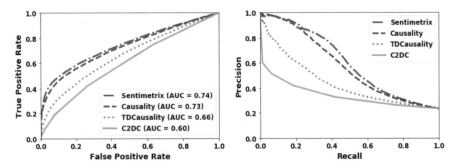

Fig. 6.3 Performance of the baseline methods on dataset $\mathscr{A}$. (Left) ROC curve. (Right) Precision-recall curve

less than 0.9 as illustrated in Fig. 6.3 (right). Note that, most of the features used in the previous bot detection work take advantage of content and network structure of users. However, this is not the case in our proposed metrics and approach.

6.4.2   Supervised Learning Approach

In this section, we describe the classification results using proposed metrics with different learning approaches. We used both datasets for this experiment. First, we use the same dataset as we used in baseline experiments ($\mathscr{A}$). Then, we use dataset $\mathscr{B}$ for comparing top methods which is 8.5 times larger than dataset $\mathscr{A}$.

Figure 6.3 (left) shows the ROC curve for different approaches. As it is shown, deep neural network achieved the highest area under the curve. Note that, the deep neural network is comparable with random forest as it is shown in Fig. 6.3 (right) on Dataset $\mathscr{A}$. The proposed approaches could improve the recall from 0.22 to 0.49 with the precision of 0.9. According to the random forest, top features are from all categories including causality metrics: ϵ_{nb}, ϵ_{wnb}, user–message graph-based metrics: user's PageRank $PR(u)$, median of retweeted message's PageRank $PS_{u,med}$, degree D_u, mean of verified users in his messages $U^v S_{u,mean}$, $\mathscr{T} S_{u,med}$, $\mathscr{T} S_{u,mean}$, median of length of the cascades $CS_{u,med}$, user graph-based metrics: weighted co-occurrence score $\mathscr{CO}^w_{u,N_u}$.

In Figs. 6.4 and 6.5, we probe the performance of the supervised approaches as well as the top two ones on the larger dataset $\mathscr{B}$, respectively. As we observe, deep neural network is able to achieve the recall of 0.48 with the precision of 0.9. It is also able to achieve AUC of 0.83 on this dataset (Fig. 6.5). It is worth to mention that we assigned higher weights to PSM accounts to deal with the data imbalance problem.

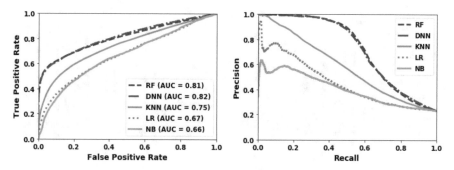

Fig. 6.4 Performance of the supervised approaches using proposed metrics on dataset $\mathscr{B}$. (Left) ROC curve. (Right) Precision-Recall curve

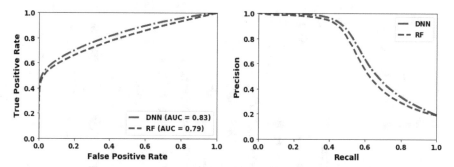

Fig. 6.5 Performance of the top two supervised approaches using proposed metrics on dataset $\mathscr{B}$. (Left) ROC curve. (Right) Precision-Recall curve

6.4.3   Self-Training Semi-Supervised Learning Approach

In this experiment, we randomly select 300 PSM and 300 normal users from dataset $\mathscr{B}$ for training and development sets and the rest of the dataset was considered as unlabeled data. We conduct two types of experiments:

WSET Algorithm In this experiment, we evaluate the self-training semi-supervised approach using Algorithm 4. In this approach, we iteratively update the training set and the termination condition is accuracy of the model on the development set. We set the parameters as $\theta_{pr} = 1$, $\alpha = 0.05$, $\theta_{tr} = 0.03$. We use a random forest classifier to train the model. The cumulative number of true positive and false positive is shown in Fig. 6.6 (left). With using 300 PSM accounts as seed set, WSET can find 29,440 PSM accounts with the precision of 0.81. Note that, we can stop the algorithm earlier.

In this case, precision varies from 0.97 to 0.81. Figure 6.6 (right) illustrates cumulative number of selected users as normal users by WSET. As shown, the number of true negatives (selected normal accounts as normal users) is 18,343 with precision of 0.93.

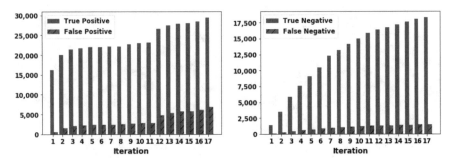

Fig. 6.6 Cumulative number of selected users using WSET Algorithm on dataset $\mathscr{B}$. (Left) PSM users. (Right) normal users

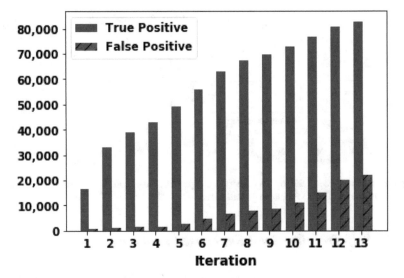

Fig. 6.7 Cumulative number of selected users as PSM accounts using supervised WSET Algorithm on dataset $\mathscr{B}$

Supervised WSET Algorithm In previous experiment, we assume that the supervisor checks accounts labeled as PSM by WSET at the end. However, this process can be done iteratively. Here, we assume that the supervisor evaluates the *PSM labeled accounts by* WSET in each iteration and verify if they are either true or false positive. Therefore, these labels along with the non-PSM labeled accounts by WSET are fed into WSET. According to our results (see Fig. 6.7), the number of true positive increases to 80,652 with the precision of more than 0.8. That is, using this approach we can increase the number of true positive PSM accounts 2.7 times.

6.5 Conclusion

In this chapter, we conducted a data-driven study on the pathogenic social media accounts. We proposed supervised and semi-supervised frameworks to detect these users. We achieved the precision of 0.9 with F1 score of 0.63 using supervised framework. In semi-supervised framework, we are able to detect more than 29K PSM users by using only 600 labeled data for training and development sets with the precision of 0.81. Our approaches identify these users without using network structure, cascade path information, content, and users' information.

References

1. H. Alvari, S. Hashemi, A. Hamzeh, Detecting overlapping communities in social networks by game theory and structural equivalence concept, in *International Conference on Artificial Intelligence and Computational Intelligence* (Springer, 2011), pp. 620–630
2. H. Alvari, A. Hajibagheri, G. Sukthankar, K. Lakkaraju, Identifying community structures in dynamic networks. Soc. Netw. Anal. Min. **6**(1), 77 (2016)
3. H. Alvari, E. Shaabani, P. Shakarian, Early identification of pathogenic social media accounts, in *ISI* (IEEE, 2018), pp. 169–174
4. G. Beigi, H. Liu, Similar but different: Exploiting users' congruity for recommendation systems, in *International Conference on Social Computing, Behavioral-Cultural Modeling and Prediction and Behavior Representation in Modeling and Simulation* (Springer, 2018), pp. 129–140
5. G. Beigi, J. Tang, H. Liu, Social science–guided feature engineering: A novel approach to signed link analysis. ACM Trans. Intell. Syst. Technol. **11**(1), (Jan. 2020)
6. F. Pedregosa, G. Varoquaux, A. Gramfort, V. Michel, B. Thirion, O. Grisel, M. Blondel, P. Prettenhofer, R. Weiss, V. Dubourg, J. Vanderplas, A. Passos, D. Cournapeau, M. Brucher, M. Perrot, E. Duchesnay, Scikit-learn: Machine learning in Python. J. Mach. Learn. Res. **12**, 2825–2830 (2011)
7. E. Shaabani, R. Guo, P. Shakarian, Detecting pathogenic social media accounts without content or network structure, in *Data Intelligence and Security (ICDIS)* (IEEE, 2018), pp. 57–64
8. E. Shaabani, A. Sadeghi-Mobarakeh, H. Alvari, P. Shakarian, An end-to-end framework to identify pathogenic social media accounts on twitter. *IEEE Conference on Data Intelligence and Security* (2019)
9. V. Subrahmanian, A. Azaria, S. Durst, V. Kagan, A. Galstyan, K. Lerman, L. Zhu, E. Ferrara, A. Flammini, F. Menczer, The DARPA twitter bot challenge. Computer **49**(6), 38–46 (2016)
10. J. Webber, I. Robinson, *A Programmatic Introduction to neo4j* (Addison-Wesley Professional, 2018)
11. X. Zhu, Semi-supervised learning literature survey. Comput. Sci. **2**(3), 4 (2006). University of Wisconsin-Madison

Chapter 7
Feature-Driven Method for Identifying Pathogenic Social Media Accounts

7.1 Introduction

The manipulation of public opinion can take many forms from fake news [22] to more subtle ones such as reinforcing specific aspects of text over others [7]. It has been observed that media aggressively exert bias in the way they report the news to sway their reader's knowledge. On the other hand, as we have learned in the previous chapters, PSM accounts are responsible for "agenda setting" and massive spread of misinformation. Understanding misinformation from account-level perspective is thus a pressing problem.

We present results from [5] which proposes an automatic feature-driven approach for detecting PSM accounts in social media. Inspired by the literature, we set out to assess PSMs from four broad perspectives: (1) causal and profile-related information, (2) source-related information (e.g., information linked via URLs), and (3) content-related information (e.g., tweets characteristics). For the causal and profile-related information, we investigate malicious signals using (1) causality analysis (i.e., if user is frequently a cause of viral cascades) [3] and (2) profile characteristics (e.g., number of followers, etc.) [16] aspects of view. For the source-related information, we explore various properties that characterize the type of information being linked to URLs (e.g., URL address, content of the associated website, etc.) [6, 10, 15, 19, 20]. Finally, for the content-related information, we examine attributes from tweets (e.g., number of hashtags, certain hashtags, etc.) posted by users [16]. This work describes the results of research conducted by Arizona State University's Global Security Initiative and Center for Strategic Communication. Research support funding was provided by the US State Department Global Engagement Center.

Our corpus comprises three different real-world Twitter datasets, from Sweden, Latvia, and United Kingdom (UK). These countries were selected to cover a range of population size and political history (former Soviet republic, neutral, founding

© The Author(s), under exclusive license to Springer Nature Switzerland AG 2021
H. Alvari et al., *Identification of Pathogenic Social Media Accounts*, SpringerBriefs in Computer Science, https://doi.org/10.1007/978-3-030-61431-7_7

member of NATO). In this study, we pose the following research questions and seek answers for them:

RQ1:   *Does incorporating information from user activities and profile character-istics help in identifying PSM accounts in social media?*

RQ2:   *What attributes could be exploited from URLs shared by users to determine whether or not they are PSMs?*

RQ3:   *Could deploying tweet-level information enhance the performance of the PSM detection approach?*

To answer **RQ1**, we first investigate different profile characteristics that could indicate suspicious behavior. Next, we also examine whether or not users who make inauthentic information go viral, are more likely to be among PSM users. By exploring **RQ2**, we figure out which characteristics of URLs and their associated websites are useful in detecting PSM users in social media. By investigating **RQ3**, we aim to examine if adding a few content-related information on tweet-level could come in handy while identifying PSMs. Our answers to the above questions lead to a feature-driven approach that uses as little as three groups of user, source, and content-related attributes to detect PSM accounts.

To summarize, this work makes the following main contributions:

- We present a feature-driven approach for detecting PSM accounts in social media. More specifically, we assess maliciousness from causal and profile-level, source-level, and content-level aspects. Our causal and profile-related information include signals in causal users (i.e., if user is frequently a cause of viral cascades) along with their profile characteristics (e.g., number of followers, etc.). For the source-related information, we explore different characteristics in URLs that users share and their associated websites (e.g., underlying themes, complexity of content, etc.). For the content-related information, we examine attributes from tweets (e.g., number of hashtags, certain hashtags, etc.) posted by users.

- We conduct a suite of experiments on three real-world Twitter datasets from different countries, using several classifiers. Using all of the attributes, we achieve average F1 scores of 0.81, 0.76, and 0.74 for Sweden, Latvian, and UK datasets, respectively. Our observations suggest the effectiveness of the proposed method in identifying PSM accounts who are more likely to manipulate public opinion in social media.

7.2   Experimental Data

We collect three real-world Twitter datasets with different number of users and tweets/retweets from three countries, Sweden, Latvia, and United Kingdom (UK). These countries were selected to cover a range of population size and political history (former Soviet republic, neutral, founding member of NATO). Description

Table 7.1 Description of the datasets used in this work

Dataset	# Tweets/retweets	# Labeled users		# Viral cascades	# URLs
		Suspended	Active		
Sweden	780,250	16,010	48,030	12,174	160,702
Latvia	323,305	10,862	32,586	1,957	76,032
UK	254,915	4,553	13,659	21,429	41,332

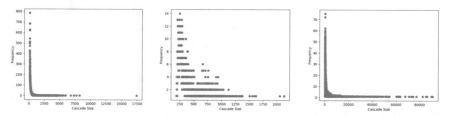

Fig. 7.1 Left to Right: Frequency plots of cascade size for Sweden, Latvia, and UK datasets

of the data is demonstrated in the Table 7.1. We use subsets of datasets from Nov 2017 to Nov 2018. Each dataset has different fields including user ID, retweet ID, hashtags, content, posting time as well as user profile information such as Twitter handles, number of followers/followees, description, location, protected, verified, etc. The tweets were collected using a predefined set of keywords and hashtags, and if they were geo-tagged in the country or user profile includes the country. We use subsets of the datasets with different number of cascades of different sizes and duration.

In our datasets, users may or may not have participated in viral cascades. We chose to use threshold $\theta = 20$ and take different number of viral cascades for each dataset with at least 20 tweets/retweets. We depict frequency plots of different cascade size for all datasets in Fig. 7.1. For brevity, we only depict cascades size greater than 100 tweets/retweets.

7.3   Identifying PSM Users

In this work, we take a machine learning approach (Fig. 7.2) to answer the research questions posed earlier in the Introduction. More specifically, we incorporate different sets of malicious behavior indicators on causal-level, account-level, source-level, and content-level to detect PSM users. In what follows, we describe each group of the attributes that will be ultimately utilized in a supervised setting to detect PSMs in social media.

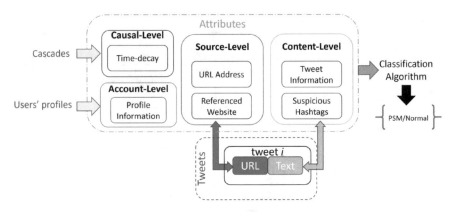

Fig. 7.2 The proposed framework for identifying PSM users. It incorporates four groups of attributes into a classification algorithm

7.3.1   Causal and Account-Level Attributes

We first set out to answer **RQ1** and understand attributes on the causal and account-level that could be exploited in order to identify PSMs in social media.

7.3.1.1   Malicious Signals in Causal Users

As stated in the previous chapters, user activity metrics are causally linked to viral cascades to the extent that malicious users who make harmful messages go viral are those with higher causality scores [3]. Accordingly, we set out to investigate if incorporating causality scores in the form of attributes in a machine learning approach, can help identify users with higher malicious behavior in social media. More specifically, we leverage the causal inference used in [1, 3, 4] to compute a vector of causality attributes for each user in our dataset. Later, these causal-based attributes will be incorporated to our final vector of attributes that will be fed into a classifier. The causal inference takes as input cascades of tweets/retweets built from the dataset. We follow the convention of [12] and assume an *action log* $\mathscr{A}$ of the form *Actions(User,Action,Time)*, which contains tuples (i, a_i, t_i) indicating that user i has performed action a_i at time t_i. For ease of exposition, we slightly abuse the notation and use the tuple (i, m, t) to indicate that user i has posted (tweeted/retweeted) message m at time t. For a given message m we define a *cascade* of actions as $\mathscr{A}_m = \{(i, m', t) \in \mathscr{A} | m' = m\}$. User i is called m-participant if there exists t_i such that $(i, m, t_i) \in \mathscr{A}$. Users who have adopted a message in the early stage of its life span are called *key users* [3].

In this work we adopt the notion of *prima facie causes* which is at the core of Suppes' theory of probabilistic causation [24] and utilize the causality metrics that are built on this theory. According to this theory, *a certain event to be recognized as a cause, must occur before the effect and must lead to an increase of the likelihood of observing the effect.* Accordingly, prima facie causal users for a given viral cascade, are key users who help make the cascade go viral. Finally, according to [3] and Chap. 4, we define causal-based attributes for each user and add them to the final representative feature vector for the given user.

In this work, we use the time-decay causal metrics introduced in [3] and Chap. 4 built on Suppes' theory [24], to build our causal-based attributes for each user. To reiterate, the first metric used in this work is $\mathscr{E}_{K\&M}$ which is computed over a given time interval I as follows:

$$\mathscr{E}_{K\&M}^{I}(i) = \frac{\sum_{j \in \mathscr{R}(i)} (\mathscr{P}_{i,j} - \mathscr{P}_{\neg i,j})}{|\mathscr{R}(i)|} \tag{7.1}$$

where $\mathscr{R}(i)$, $\mathscr{P}_{i,j}$, and $\mathscr{P}_{\neg i,j}$ are now defined over I. This metric estimates the causality score of user i in making a message *viral*, by taking the average of $\mathscr{P}_{i,j} - \mathscr{P}_{\neg i,j}$ over $\mathbf{R}(i)$. The intuition here is that user i is more likely to be a cause of message m to become viral than user j, if $\mathscr{P}_{i,j} - \mathscr{P}_{\neg i,j} > 0$. This metric cannot spot all PSMs, hence another metric is defined, namely relative likelihood causality $\mathscr{E}_{rel}$. This metric works by assessing relative difference between $\mathscr{P}_{i,j}$, and $\mathscr{P}_{\neg i,j}$:

$$\mathscr{E}_{rel}^{I}(i) = \frac{\mathscr{S}(i, j)}{|\mathscr{R}(i)|} \tag{7.2}$$

where $\mathscr{S}(i, j)$ is defined as follows and α is infinitesimal:

$$\mathscr{S}(i, j) = \begin{cases} \frac{\mathscr{P}_{i,j}}{\mathscr{P}_{\neg i,j} + \alpha} - 1, & \mathscr{P}_{i,j} > \mathscr{P}_{\neg i,j} \\ 1 - \frac{\mathscr{P}_{\neg i,j}}{\mathscr{P}_{i,j}}, & \mathscr{P}_{i,j} \le \mathscr{P}_{\neg i,j} \end{cases} \tag{7.3}$$

Two other neighborhood-based metrics were also defined in [3], first of which is computed as

$$\mathscr{E}_{nb}^{I}(j) = \frac{\sum_{i \in \mathscr{Q}(j)} \mathscr{E}_{K\&M}^{I}(i)}{|\mathscr{Q}(j)|} \tag{7.4}$$

where $\mathscr{Q}(j) = \{i | j \in \mathscr{R}(i)\}$ is the set of all users that user j belongs to their related users sets. Similarly, the second metric is the weighted version of the above metric and is called weighted neighborhood-based causality and is calculated as

$$\mathscr{E}_{wnb}^{I}(j) = \frac{\sum_{i \in \mathscr{Q}(j)} w_i \times \mathscr{E}_{K\&M}^{I}(i)}{\sum_{i \in \mathscr{Q}(j)} w_i} \tag{7.5}$$

The aim of this metric is to capture different impacts that users in $Q(j)$ might have on user j.

Finally, the causal metrics discussed above will be utilized to compute causality features for each user. These features are added to our final vector representation for each user.

The final set of features is in the following generic form ξ_k^I [3]:

$$\xi_k^I(i) = \frac{1}{|\mathcal{T}|} \sum_{t' \in \mathcal{T}} e^{-\sigma(t-t')} \times \mathcal{E}_k^{\Delta}(i) \tag{7.6}$$

Here, $k \in \{K\&M, rel, nb, wnb\}$, σ is a scaling parameter of the exponential decay function, $\mathcal{T} = \{t' | t' = t_0 + j \times \delta, j \in \mathbb{N} \wedge t' \le t - \delta\}$ is a sequence of sliding-time windows, and δ is a small fixed amount of time, which is used as the length of each sliding-time window $\Delta = [t' - \delta, t']$.

7.3.1.2   Malicious Signals in Profile Characteristics

Having defined our causality-based attributes, we now describe our next set of user-based features. Specifically, for each user, we collect account-level features and add them to the final feature vector for that user. We follow the work of [16] and compute the following 10 features from users' profiles: *Statuses Count, Followers Count, Friends Count, Favorites Count, Listed Count, Default Profile, Geo Enables, Profile Uses Background Image, Verified, Protected.* Prior research has shown promising results using this small set of features [11] with far less number of features than the established bot detection approach, namely, Botometer which uses over 1,500 features. Accordingly, we extend the final feature vector representation of each user by adding these 10 features.

7.3.2   Source-Level Attributes

Here, we seek an answer to **RQ2** and examine malicious behavior from the source-level perspective. Previous research has demonstrated the differences between normal and PSM users in terms of their shared URLs [2] and their impact on creating subsequent events in future (see also Chap. 2). We thus follow the same Hawkes processes procedure described in Chap. 2 and compute the infectivity matrices for the Sweden, Latvia, and UK datasets. The matrices are depicted in Fig. 7.3. Similarly, we observe clear distinctions between PSM and normal users' behaviors in terms of their shared URLs. Specifically, URLs shared by PSM accounts more likely trigger subsequent events (i.e., future adoptions of URLs) when coming from alternative news sources. This is in contrast to the URLs shared by normal users which either trigger subsequent events on mainstream news outlets or social media platforms. Following our observations, we now take URLs posted by users as source-related information that could be used in our PSM user detection approach.

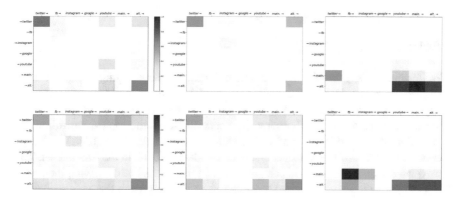

Fig. 7.3 (Top) Left to Right: Estimated infectivity matrices for PSM accounts in Sweden, Latvia, and UK Datasets. (Bottom) From Left to Right: Estimated infectivity matrices for normal users in Sweden, Latvia, and UK Datasets

Specifically, we set out to understand several characteristics of each URL from two broad perspectives: (1) URL address and (2) content collected from the website it has referenced.

7.3.2.1   URL Address

Far-Right and Pro-Russian URLs

Here, we examine if the given URL refers to the known far-right websites.[1] We further note that each user may have posted multiple URLs posted in our data. To account for that, we compute the average of these attribute values for each user. Ultimately, this list leads to a vector of 5 values for each URL shared by each user in our dataset. We leave examining other malicious websites to future work.

Domain Extensions

Previous research on assessing news articles credibility suggests looking at their URLs [6] to examine if they contain features such as whether a website contains the *http* or *https* prefixes, or *.gov*, *.co*, and *.com* domain extensions. Likewise, we investigate if the URLs in our dataset contain any of these 5 features by counting the number of times each URL triggers one of these attributes and taking the average if user has shared multiple of such URLs. This additional attribute vector will be added to the final attribute vector for each user.

---

[1] In this work we used the following: https://voiceofeurope.com/, https://newsvoice.se/, https://nyadagbladet.se/, https://www.friatider.se/, or the pro-Russian website https://ok.ru/.

7.3.2.2 Referenced Website Content

Topics

We further investigate whether or not incorporating the underlying topics or themes learned from the text of the websites, could help us to build a more accurate approach to identify malicious activity. More specifically, we first set out to extract the content from each URL shared by users. To learn the topics, we follow the procedure described in [20] and train Latent Dirichlet Allocation (LDA) [8] on the crawled contents of the websites associated with each URL in the training set. This way, we obtain a fine-grained categorization of the URLs by identifying their most representative topics as opposed to a coarser-grained approach that uses predefined categories (e.g., sports, etc.). Using LDA also allows for uncovering the hidden thematic structure in the training data. Furthermore, we rely on the document-topic distribution given by the LDA (here each document is seen as a mixture of topics) to distinguish normal users from highly biased users. After training LDA, we treat each new document and measure their association with each of the K topics discovered by LDA. We empirically found $K = 25$ to work well in our dataset. Thus, each document is now treated as a vector of 25 probabilistic values provided by LDA's document-topic distribution—this feature space will be added to the final set of the features built so far. Finally, note that for users with more than one URL, we take the average of different probabilistic feature vectors.

Has Quote

Social science research has shown that news agencies seek to make a piece of information more noticeable, meaningful, and memorable to the audience [10]. This increases the chance of shifting believes and perceptions. One way to increase salience of a piece of information is emphasizing it by selecting particular facts, concepts, and quotes that match the targeted goals [9, 10, 21]. We thus check the existence of quotes within the referenced website content as an indicator of malicious behavior—this results in a single binary feature. Each user may post more than one URL. To account for this, we take the average values of this feature for each user. We observe that the PSM users' mean scores for this feature are 0.04 (Sweden), 0.05 (Latvia), and 0.04 (UK). Normal users have mean scores of 0.05 (Sweden), 0.05 (Latvia), and 0.03 (UK). We also deploy two-tailed two-sample t-test with the null hypothesis that value of this feature is not significantly different between normal and PSM accounts. Table. 7.2 summarizes the p-values for this test with significance level $\alpha = 0.01$. Results show that the null hypothesis could not be rejected. However, we still include this feature to see whether or not it helps in identifying PSMs in practice.

Table 7.2 Results of p-values at significance level $\alpha = 0.01$. The null hypotheses for complexity and readability tests are refuted

Feature	Sweden	Latvia	UK
Has Quote	0.29	0.36	0.32
Complexity	4.95e−50	3.23e−07	6.12e−08
Readability	5.56e−27	4.2e−03	1.9e−14

Complexity

Research has shown that complexity of the given text could be different for malicious and normal users [19]. We thus use complexity feature to see whether or not it aids the classifier in finding users who create and share malicious content. We follow the same approach as in [19] and approximate the complexity of reference website content as follows:

$$\text{complexity} = \frac{\text{number of unique part-of-speech tags}}{\text{number of words in the text}} \tag{7.7}$$

The higher this score is, the more complex the given context is. Surprisingly, our initial analysis show that mean of complexity score of website content by PSMs are 0.53 (Sweden), 0.54 (Latvia), and 0.51 (UK) while mean of complexity score of website contents shared by normal users are 0.46 (Sweden), 0.51 (Latvia), and 0.48 (UK). This shows contents shared by PSMs have higher complexity than those shared by normal users. We also deploy one-tail two-sample t-test with the null hypothesis that content of URLs shared by normal are more complex than those shared by PSMs. Table 7.2 summarizes the p-values showing that the null hypothesis was rejected at significance level $\alpha = 0.01$. This indicates that content of websites referenced by PSM users are more complex than those shared by normal users.

Readability

According to [14], readability of a given context can affect engagement of the individuals with the given piece of information. Therefore, readability of the referenced website content is another important feature which could be useful in distinguishing PSMs and normal users. We hypothesize that PSM users may share information with higher readability to increase the chance of transferring the concept and creating malicious content. We use Flesch–Kincaid reading-ease test [15] on the text of the provided URLs. The mean readability scores are 61.16 (Sweden), 62.98 (Latvia), 59.08 (UK) for PSMs and 55.44 (Sweden), 56.79 (Latvia), 55.35 (UK) for other normal users. The higher the score is, the more readable the text is. We also deploy one-tail two-sample t-test with the null hypothesis that content of URLs shared by normal users are more readable than those shared by PSMs. Table 7.2 summarizes the p-values indicating that the null hypothesis was rejected at significance level $\alpha = 0.01$.

Table 7.3 Top selected bigrams for each country

Data	Bigrams
Sweden	Asylum seeker, birthright citizenship, court justice, European commission, European Union (EU), European parliament, kill people, migrant caravan, national security, Russian military, school shooting, sexually assault, United Nations, white supremacist, police officer
Latvia	Baltic exchange, Baltic security, battlefield revolution, cyber security, depository Estonia, Estonia Latvia, European parliament, European commission, European Union (EU), human rights, Latvian government, nasdaq Baltic, national security, Saeima election, Vladimir Putin
UK	Court appeal, cosmic diplomacy, defence police, depression anxiety, diplomacy ambiguity, European Union (EU), human rights, Jewish community, police officer, police federation, political party, rebel medium, sexually liberate, support group, would attack

These results show that the content of URLs shared by PSM accounts are more complex yet more readable than those shared by normal users. Therefore, these two features, complexity and readability, could be a good indicator to distinguish between normal and PSMs.

Unigrams/Bigrams

We use TF-IDF weighting for extracted word-level unigrams and bigrams. This feature gives us both importance of a term in the given context (i.e., term frequency) and term's importance considering the whole corpus. We remove stop words and select top 20 frequent unigrams/bigrams as the final set of features for this group. Using TF-IDF weighting helps to identify piece of information that is focusing on aspects not emphasized by others. For brevity, we only demonstrate top bigrams in Table 7.3.

Domain Expertise

The presence of signal words (e.g., specific frames or keywords) could be indicator of existence of malicious behavior in the text. In this work, we hired human coders and trained them based on our codebook[2] in order to provide signal words that can help identify suspicious behavior. We use the following framing categories: *Anti-immigrant, Crime rampant, Government, Anti-EU/NATO, Russia-ally, Crimea, Discrimination, Fascism.* For each country and each category, we have a list of

---

[2] A codebook is survey research approach to provide a guide for framing categories and coding responses to the categories definitions.

Table 7.4 Examples of the keywords used in this study

Data	Keywords
Sweden	No-go zones, violence overwhelmed, police negligence, Nato obsolete, bilateral cooperation, blighted areas, increase reported rapes, close police station, EU hypocrisy, anti-immigrant, fatal shootings, badly Sweden, Nato airstrikes
Latvia	Brussels silent, norms international law, bureaucrats, lack trust EU, based universal principles, Russia borders, anti Nato, purely political, European bureaucrats, silence Brussels Washington, rampant, harsh statements concerning, values Brussels silent
UK	Brexit, Theresa May, stop Brexit, hard Brexit, post Brexit, leave, referendum, Brexitshambles

corresponding keywords. We have illustrated examples of the keywords used in this study in Table 7.4.

7.3.3   Content-Level Attributes

In this section, we aim to understand **RQ3** by incorporating a few more attributes from the content-level information that could be used to enhance the performance of the PSM user detection. For the content-level information, we only rely on the tweets posted by each user in our dataset.

7.3.3.1   Malicious Signals in Tweet-Level Information

We use the following 6 attributes extracted from each tweet [16]: *retweet count, reply count, favorite count, number of hashtags, number of URLs, number of mentions*. If the user has tweeted more than once, we take the average of these features.

7.3.3.2   Malicious Signals in Suspicious Hashtags

We further investigate if the given tweet aims to push propaganda using any of the following suspicious hashtags identified by our human coders. For Sweden, we use *#Swedistan, #Swexit, #sd (far right group), #SoldiersofOdin, #NOGO-Zones*. For Latvia, we use *#RussiaCountryFake, #BrexitChaos, #BrexitVote, #Soviet, #RussiaAttacksUkraine*. For UK, we use *#StopBrexit, #BrexitBetrayal, #StopBrexit-SaveBritain, #StandUp4Brexit, #LeaveEU*. Similar to the previous attributes, for the users who have posted more than one tweet with these hashtags, we compute the average of the corresponding values. We leave examining other suspicious hashtags to future work.

7.3.4 Feature-Driven Approach

Having described the attributes (Table 7.5) used in this work, we now feed them into a supervised classification algorithm to detect PSM users (Fig. 7.2). In more details, we feed the profile information and tweets into the different components of the proposed approach. For the causal and account-related information, we require both of the profile characteristics and tweets. We need tweets to build viral cascades and finally compute causality scores for different users. Each cascade contains tuples (i, m, t) indicating that user i has posted (tweeted/retweeted) the corresponding message m at time t. Given the cascades, causality features are computed for each user i based on her activity log in our dataset. For the source-level information, we only need to extract URLs from tweets. These URLs are either directly used to compute attributes or to collect the content from the websites to which they have referenced. For the content-related information, we only need tweets in order to compute the content-level attributes. Finally, for each user, we fuse all attributes into a feature vector representation and feed them into a classifier.

Table 7.5 Different groups of features used in this work

	Feature	Definition	# Feat.
Causal	Time-decay	Attributes computed using causality-based metrics	4
Account	Profile-based	*Statuses Count, Followers Count, Friends Count, Favorites Count, Listed Count, Default Profile, Geo Enables, Profile Uses Background Image, Verified, Protected*	10
Source	Websites	Presence of far-right and pro-Russian websites	5
	Domains	Existence of *http* or *https* prefixes, or .*gov*, .*co*, and .*com* domain extensions	5
	Topics	Features computed by comparing the listing against the learned topic distribution	25
	Has Quote	Single binary feature that shows whether the content of shared URLs contains quote or not	1
	Complexity	Complexity of content of shared URLs by users	1
	Readability	Readability of content of shared URLs by users	1
	Unigram	TF-IDF scores of highly frequent word-level unigrams extracted from content of URLs shared by users	20
	Bigram	TF-IDF scores of highly frequent word-level bigrams extracted from content of URLs shared by users	20
	Expertise	Presence of signal keywords provided by our coders	8
Content	Tweet-based	*Retweet count, reply count, favorite count, number of hashtags, number of URLs, number of mentions*	6
	Hashtags	Presence of suspicious hashtags	5

7.4 Experiments

In this section, we conduct experiments on three real-world Twitter datasets to gauge the effectiveness of the proposed approach. In particular, we compare the results of several classifiers and baseline methods. Note for all methods, we only report results when their best settings are used.

- **Ensemble Classifiers**

  - **Gradient Boosting Decision Tree (GBDT)** We train a Gradient Boosting Decision Tree classifier using the described features. We set the number of estimators as 200. Learning rate was set to the default value of 0.1.
  - **Random Forest (RF)** We train a Random Forest classifier using the features described. We use 200 estimators and entropy as the criterion.
  - **AdaBoost** We train an AdaBoost classifier using the described features. The number of estimators was set to 200 and we also set the learning rate to 0.01.

- **Discriminative Classifiers**

  - **Logistic Regression (LR)** We train a Logistic Regression using $l2$ penalty. We also set the parameter $C = 1$ (the inverse of regularization strength) and tolerance for stopping criteria to 0.01.
  - **Decision Tree (DT)** We train a Decision Tree classifier using the features. We did not tune any specific parameter.
  - **Support Vector Machines (SVM)** We use a linear SVM using the attributes described in the previous section. We set the tolerance for stopping criteria to 0.001 and the penalty parameter $C = 1$.

- **Generative Classifiers**

  - **Naive Bayes (NB)** We train a Multinomial Naive Bayes which has shown promising results for text classification problems [17]. We did not tune any specific parameter for this classifier

- **Baselines**

  - **Long Short-Term Memory (LSTM) [16]** The word-level LSTM approach here is similar to the deep neural network models used for sequential word predictions. We adapt the neural network to a sequence classification problem where the inputs are the vector of words in each tweet and the output is the predicted label of the tweet. We first use the word2vec [18] embeddings which are trained jointly with the classification model. We use a single LSTM layer of 50 units on the textual content, followed by the loss layer which computes the cross entropy loss used to optimize the model.
  - **Account-Level (AL) + Random Forest [16]** This approach uses the following features of the user profiles: *Statuses Count, Followers Count, Friends Count, Favorites Count, Listed Count, Default Profile, Geo Enables, Profile Uses Background Image, Verified, Protected.* We chose this method

over Botometer [25] as it achieved comparable results with far less number of features ([25] uses over 1,500 features) (see also [11]). According to [16], we report the best results when Random Forest (RF) is used.

- **Tweet-Level (TL) + Random Forest [16].** Similar to the previous baseline, this method uses only a handful of features extracted from tweets: *retweet count, reply count, favorite count, number of hashtags, number of URLs, number of mentions.* Likewise, we use RF as the classification algorithm.

7.4.1 Results and Discussion

All experiments were implemented in Python 2.7x and run on a machine equipped with an Intel(R) Xeon(R) CPU of 3.50 GHz with 200 GB of RAM running Linux. We use tenfold cross-validation as follows. We first divide the entire set of training instances into 10 different sets of equal sizes. Each time, we hold one set out for validation. This procedure is performed for all approaches and all datasets for the sake of fair comparison. Finally, we report the average of 10 different runs, using F1-macro and F1-score (only for PSM users) evaluation metrics and all features in Table 7.6.

7.4.1.1   Performance Evaluation

For any approach that requires special tuning of parameters, we conducted grid search to choose the best set of parameters. Also, for LSTM, we preprocess the individual tweets in line with the steps mentioned in [23]. We use word vectors of dimensions 100 and deploy the skip-gram technique for obtaining the word vectors where the input is the target word, while the outputs are the words surrounding the

Table 7.6 Performance comparison on different datasets using all features

Classifier	Sweden		Latvia		UK	
	F1-macro	F1-score	F1-macro	F1-score	F1-macro	F1-score
GBDT	**0.80**	**0.81**	**0.76**	**0.76**	**0.73**	**0.74**
RF	0.79	0.79	0.75	0.75	0.70	0.71
AdaBoost	0.78	0.79	0.73	0.74	0.69	0.70
LR	0.75	0.75	0.74	0.74	0.71	0.72
DT	0.69	0.69	0.71	0.71	0.69	0.69
SVM	0.73	0.74	0.73	0.70	0.72	0.70
NB	0.71	0.71	0.65	0.67	0.66	0.67
LSTM	0.60	0.62	0.58	0.65	0.36	0.43
AL (RF)	0.64	0.64	0.63	0.64	0.64	0.65
TL (RF)	0.50	0.51	0.50	0.51	0.49	0.50

target words. To model the tweet content in a manner that uses it to predict whether an account is biased or not, we used LSTM models [13]. For the LSTM architecture, we use the first 20 words in the tokenized text of each tweet and use padding in situations where the number of tokens in a tweet is less than 20. We use 30 units in the LSTM architecture (many to one). The output of the LSTM layer is fed to a dense layer of 32 units with ReLU activations. We add dropout regularization following this layer to avoid overfitting and the output is then fed to a dense layer which outputs the category of the tweets.

Observations Overall, we make the following observations:

- In general, results from different classifiers compared to the baselines demonstrate the effectiveness of the described attributes in identifying PSM users in social media. Thus, the answers to the research questions **RQ1–RQ3** are all positive, i.e., we could exploit attributes from user activities and profile characteristics, source and content-related information for identifying PSM users in social media. More specifically, for **RQ1**, we investigate different profile characteristics that could indicate suspicious behavior. We also examine whether or not users who make inauthentic information go viral, are more likely to be among PSM users. By answering **RQ2**, we figure out which characteristics of URLs and their associated websites are useful in detecting PSM users in social media. By investigating **RQ3**, we examine if adding a few content-related information on tweet-level could come in handy while identifying PSMs. Our answers to the above questions lead to a feature-driven approach that uses as little as three groups of user, source, and content-related attributes to detect PSM accounts.
- Ensemble classifiers using the described features, outperform all other classifiers and baselines. Among the ensemble classifiers, Gradient Boosting Decision Trees classifier achieves the best results in terms of both F1-macro and F1-score metrics.
- Among the discriminative classifiers, linear Support Vector Machines classifier marginally beats Logistic Regression. Decision Tree classifier achieves the worst results in this category.
- Overall, Decision Tree and Naive Bayes classifiers achieve the worst performance among all classifiers.
- For LSTM, we achieve slightly poor performance than the logistic regression classifier. One reason behind the poor performance of the classifier is the lack of trained word embeddings suited to our dataset. Also, the poor performance might suggest that the sequential nature of the texts might not be very helpful for the task of PSM users detection.
- Overall, results on Sweden data demonstrate better performances achieved using the attributes. One reason behind this might be the size of data and higher number of PSMs in Sweden data compared to others. This could also indicate that PSMs in Latvia and UK data are more sophisticated.

Table 7.7 Feature importance on different datasets

Feature	Sweden	Latvia	UK
Causal	0.64	0.62	0.61
Account	0.62	0.61	0.59
Content	0.45	0.43	0.40
Source	0.73	0.70	0.68
All \ Source	0.71	0.65	0.63
All \ Causal	0.73	0.67	0.62
All \ Account	0.76	0.70	0.69
All \ Content	0.79	0.73	0.72
All	0.81	0.76	0.74

7.4.1.2   Feature Importance Analysis

We further conduct feature import analysis to investigate what feature group contributes the most to the performance of the proposed approach. More specifically, we use GBDT and perform different 10-fold cross validations using each feature group. We report the F1-score results in Table 7.7. According to our observations, we conclude that the most significant and less significant feature groups are *source-related* and *content-related* attributes, respectively. We also perform feature ablation test by taking out a single feature group at a time from the rest. We observe that eliminating content-related attributes has the least impact on the performance, while taking out source-related attributes deteriorates the performance drastically. One final note though is, despite the effectiveness of the attributes from the user-level information, they may not be always available or we may not always know the suspicious sources beforehand for the task at hand. This further demonstrates the effectiveness of the causal-related features extracted from users' activities for identifying PSM users and thus confirms the observations in the previous chapters.

7.5   Conclusion

We present an automatic feature-driven approach for identifying PSM accounts in pro-Russian social media. In particular, we assess the malicious behavior from four broad perspectives: (1) causal, (2) account, (3) source, and (4) content-related information. For the first two groups, we investigate malicious signals using (1) causality analysis (i.e., if user is frequently a cause of viral cascades) and (2) profile characteristics (e.g., number of followers, etc.) aspects of view. For the source-related information, we explore various properties that characterize the type of information being linked to URLs (e.g., URL address, content of the associated website, etc.). Finally, for the content-related information, we examine attributes from tweets (e.g., number of hashtags, certain hashtags, etc.).

References

1. H. Alvari, P. Shakarian, Causal inference for early detection of pathogenic social media accounts. Preprint (2018). arXiv:1806.09787
2. H. Alvari, P. Shakarian, Hawkes process for understanding the influence of pathogenic social media accounts, in *2019 2nd International Conference on Data Intelligence and Security (ICDIS)*, pp. 36–42 (June 2019)
3. H. Alvari, E. Shaabani, P. Shakarian, Early identification of pathogenic social media accounts. *IEEE Intelligent and Security Informatics* (2018). arXiv:1809.09331
4. H. Alvari, E. Shaabani, S. Sarkar, G. Beigi, P. Shakarian, Less is more: Semi-supervised causal inference for detecting pathogenic users in social media, in *Companion Proceedings of The 2019 World Wide Web Conference*, WWW '19 (Association for Computing Machinery, New York, NY, USA, 2019), pp. 154–161
5. H. Alvari, G. Beigi, S. Sarkar, S. W. Ruston, S. R. Corman, H. Davulcu, P. Shakarian, A feature-driven approach for identifying pathogenic social media accounts. Preprint (2020). arXiv:2001.04624
6. R. Baly, G. Karadzhov, D. Alexandrov, J. Glass, P. Nakov, Predicting factuality of reporting and bias of news media sources. Preprint (2018). arXiv:1810.01765
7. D.P. Baron, Persistent media bias. J. Public Econ. **90**(1-2), 1–36 (2006)
8. D.M. Blei, A.Y. Ng, M.I. Jordan, Latent Dirichlet allocation. J. Mach. Learn. Res. **3**, 993–1022 (2003)
9. S. DellaVigna, E. Kaplan, The Fox News effect: Media bias and voting. Q. J. Econ. **122**(3), 1187–1234 (2007)
10. R.M. Entman, Framing: Toward clarification of a fractured paradigm. J. Commun. **43**(4), 51–58 (1993)
11. E. Ferrara, O. Varol, C. Davis, F. Menczer, A. Flammini, The rise of social bots. Commun. ACM **59**(7), 96–104 (2016)
12. A. Goyal, F. Bonchi, L.V. Lakshmanan, Learning influence probabilities in social networks, in *WSDM* (2010)
13. S. Hochreiter, J. Schmidhuber, Long short-term memory. Neural Computation **9**(8), 1735–1780 (1997)
14. B.D. Horne, S. Adali, This just in: Fake news packs a lot in title, uses simpler, repetitive content in text body, more similar to satire than real news, in *Eleventh International AAAI Conference on Web and Social Media* (2017)
15. J.P. Kincaid, R.P. Fishburne Jr., R.L. Rogers, B.S. Chissom, Derivation of new readability formulas (automated readability index, fog count and Flesch reading ease formula) for navy enlisted personnel (1975)
16. S. Kudugunta, E. Ferrara, Deep neural networks for bot detection. Preprint (2018). arXiv:1802.04289
17. C. Manning, R. Prabhakar, S. Hinrich, *Introduction to Information Retrieval*, vol. 1 (Cambridge University Press, Cambridge, 2008)
18. T. Mikolov, I. Sutskever, K. Chen, G.S. Corrado, J. Dean, Distributed representations of words and phrases and their compositionality, in *Advances in Neural Information Processing Systems*, pp. 3111–3119 (2013)
19. F. Morstatter, L. Wu, U. Yavanoglu, S.R. Corman, H. Liu, Identifying framing bias in online news. ACM Trans. Soc. Comput. **1**(2), 5 (2018)
20. T.M. Phuong et al., Gender prediction using browsing history, in *Knowledge and Systems Engineering* (Springer, 2014), pp. 271–283
21. D.A. Scheufele, D. Tewksbury, Framing, agenda setting, and priming: The evolution of three media effects models. J. Commun. **57**(1), 9–20 (2006)
22. C. Shao, G.L. Ciampaglia, O. Varol, A. Flammini, F. Menczer, The spread of fake news by social bots. Preprint (2017). arXiv:1707.07592

23. A.B. Soliman, K. Eissa, S.R. El-Beltagy, Aravec: A set of Arabic word embedding models for use in Arabic NLP. Procedia Comput. Sci. **117**, 256–265 (2017)
24. P. Suppes, A probabilistic theory of causality (1970)
25. O. Varol, E. Ferrara, C.A. Davis, F. Menczer, A. Flammini, Online human-bot interactions: Detection, estimation, and characterization, in *ICWSM* (2017)

Chapter 8
Conclusion

In this book, we presented results of the efforts to detect "Pathogenic Social Media (PSM)" accounts who are responsible for manipulating public opinion and political events. There are many challenges in the area of PSM accounts detection. In Chaps. 3 and 4, standard and time-decay probabilistic causal metrics were proposed to distinguish PSM from normal users within a short time around their activity. In Chap. 4, we investigated whether or not causality scores of PSM users within same communities are higher than those across different communities. Furthermore, as available data for training automatic approaches for detecting PSM users are usually either highly imbalanced or comprise insufficient labeled data, in Chaps. 5 and 6, we proposed semi-supervised approaches for detecting PSMs that utilize unlabeled data to compensate for the lack of sufficient labeled data. In Chap. 7, we observed that PSMs would deploy techniques to generate diverse information to make their posts look more natural. We utilize several metrics to approximate the complexity and readability of content shared online by PSMs and normal users. Finally, in Chap. 7, we took a closer look at the differences between malicious and normal behavior in terms of the posted URLs by different types of users. We leveraged several characteristics of URLs as source-level information along with other attributes in a supervised setting for detecting PSMs.

In future we plan to incorporate time-related attributes from point processes, time-series, and LSTM, to support more real-time detection of PSM accounts. Our future plans also include investigating other forms of causality inferences and regularization terms to seek if we can further improve the classification performance. Another direction for future work would be reducing false positives which is too costly for social media. Finally, we would also like to present methods that can distinguish between different types of PSM users.

© The Author(s), under exclusive license to Springer Nature Switzerland AG 2021
H. Alvari et al., *Identification of Pathogenic Social Media Accounts*, SpringerBriefs
in Computer Science, https://doi.org/10.1007/978-3-030-61431-7_8

SPRINGER BRIEFS IN COMPUTER SCIENCE

Hamidreza Alvari · Elham Shaabani · Paulo Shakarian

Identification of Pathogenic Social Media Accounts
From Data to Intelligence to Prediction

ISBN 978-3-030-61430-0

António Gusmão
Nuno Horta
Nuno Lourenço
Ricardo Martins

Analog IC Placement Generation via Neural Networks from Unlabeled Data